MEI STRUCTURED MATHEMATICS

SECOND EDITION

Mechanics 1

John Berry
Pat Bryden
Ted Graham
Roger Porkess

Series Editor: Roger Porkess

Hodder & Stoughton

A MEMBER OF THE HODDER HEADLINE GROUP

Acknowledgements

We are grateful to the following companies, institutions and individuals who have given permission to reproduce photographs in this book. Every effort has been made to trace and acknowledge ownership of copyright. The publishers will be glad to make suitable arrangements with any copyright holders whom it has not been possible to contact.

Ruth Nossek (page 1); Andrew Ward/Life File (page 5); ATOC 1998 (Figure 2.1); David Cumming; Eye Ubiquitous CORBIS (page 22); Topham Picturepoint – Korea, 1950 (page 37); Action-Plus Photographic (page 42); Jeremy Hoare/Life File – commuter train in Korea (page 44), oil tanker (page 61); NASA (page 50); N.P.G., London – Sir Isaac Newton (page 56); Mehau Kulyk/Science Photo Library (page 74); John Cox/Life File (page 78, left), Graham Buchan/Life File (page 78, centre), B & C Alexander (page 78, right); John Greim/Science Photo Library (page 101); Action-Plus Photographic (page 123); Emma Lee/Life File – cable car, Singapore (page 130); Jennie Woodcock; Reflections PhotoLibrary/CORBIS (page 147).

OCR, AQA and Edexcel accept no responsibility whatsoever for the accuracy or method of working in the answers given.

Orders: please contact Bookpoint Ltd, 78 Milton Park, Abingdon, Oxon OX14 4TD. Telephone: (44) 01235 827720, Fax: (44) 01235 400454. Lines are open from 9.00–6.00, Monday to Saturday, with a 24 hour message answering service. Email address: orders@bookpoint.co.uk

British Library Cataloguing in Publication Data
A catalogue record for this title is available from The British Library

ISBN 0 340 771909

First published 1993
Second edition published 2000
Impression number   10 9 8 7 6 5 4 3 2
Year   2005 2004 2003 2002 2001 2000

Copyright © 1993, 2000, John Berry, Pat Bryden, Ted Graham, David Holland, Roger Porkess

Typeset by Tech-Set Ltd, Gateshead, Tyne & Wear.
Printed in Great Britain for Hodder & Stoughton Educational, a division of Hodder Headline Plc, 338 Euston Road, London NW1 3BH by J.W. Arrowsmiths Ltd, Bristol.

MEI Structured Mathematics

Mathematics is not only a beautiful and exciting subject in its own right but also one that underpins many other branches of learning. It is consequently fundamental to the success of a modern economy.

MEI Structured Mathematics is designed to increase substantially the number of people taking the subject post-GCSE, by making it accessible, interesting and relevant to a wide range of students.

It is a credit accumulation scheme based on 45 hour modules which may be taken individually or aggregated to give Advanced Subsidiary (AS) and Advanced GCSE (A Level) qualifications in Mathematics, Further Mathematics and related subjects (like Statistics). The modules may also be used to obtain credit towards other types of qualification.

The course is examined by OCR (previously the Oxford and Cambridge Schools Examination Board) with examinations held in January and June each year.

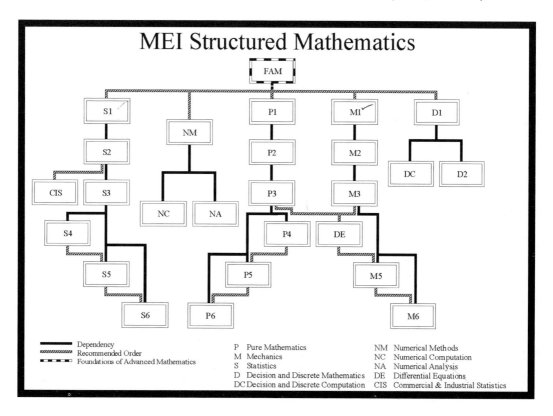

This is one of the series of books written to support the course. Its position within the whole scheme can be seen in the diagram above.

Mathematics in Education and Industry is a curriculum development body which aims to promote the links between Education and Industry in Mathematics at secondary school level, and to produce relevant examination and teaching syllabuses and support material. Since its foundation in the 1960s, MEI has provided syllabuses for GCSE (or O Level), Additional Mathematics and A Level.

For more information about MEI Structured Mathematics or other syllabuses and materials, write to MEI Office, Albion House, Market Place, Westbury, Wiltshire, BA13 3DE.

Introduction

This is the first in a series of books to support the Mechanics modules in MEI Structured Mathematics. They are also suitable for other courses in the subject. Throughout the series emphasis is placed on understanding the basic principles of mechanics and the process of modelling the real world, rather than on mere routine calculations.

In this book you meet the basic concepts, laws of motion and force, vector techniques and the modelling cycle. Some examples of everyday applications are covered in the worked examples in the text, many more in the various exercises. Working through these exercises is an important part of learning the subject. Not only will it help you to appreciate the wide variety of situations that can be analysed using mathematics, it will also build your confidence in applying the techniques you have learnt. You will find a number of quite short exercises placed in the middle of topics; these are meant to be done as you meet them to ensure that you have fully understood the ideas so far.

Mechanics is not just a pen-and-paper subject. It is about modelling the real world and this involves observing what is going on around you. This book includes a number of simple experiments and investigations for you to carry out. Make sure you do so; they will really help you.

We have used S.I. units throughout the book but have from time to time included examples using other common units for example mph and knots.

This is the second edition of *Mechanics 1* in this series and is essentially a new book. However we have used some sections of the previous text and quite a number of the questions. We would like to acknowledge and thank John Berry, Ted Graham and David Holland for this contribution to the book. We would also like to thank all those who have looked through early drafts of the text and made helpful suggestions, particularly Robin Grayson and Ruth Stanier. Finally we would like to thank the various examination boards who have given permission for their past questions to be used in the exercises.

Pat Bryden, Roger Porkess

Contents

1 Motion

The whole burden of philosophy seems to consist in this – from the phenomena of motions to investigate the forces of nature.

Issac Newton

The language of motion

Throw a small object such as a marble straight up in the air and think about the words you could use to describe its motion from the instant just after it leaves your hand to the instant just before it hits the floor. Some of your words might involve the idea of direction. Other words might be to do with the position of the marble, its speed or whether it is slowing down or speeding up. Underlying many of these is time.

Direction

The marble moves as it does because of the gravitational pull of the earth. We understand directional words such as up and down because we experience this pull towards the centre of the earth all the time. The *vertical* direction is along the line towards or away from the centre of the earth.

In mathematics a quantity which has only size, or magnitude, is called a *scalar*. One which has both magnitude and a direction in space is called a *vector*.

Distance, position and displacement

The total *distance* travelled by the marble at any time does not depend on its direction. It is a scalar quantity.

Position and displacement are two vectors related to distance, they have direction as well as magnitude. Here their direction is up or down and you decide which of these is positive. When up is taken to be positive, down is negative.

The *position* of the marble is then its distance above a fixed origin, for example the distance above the place it first left your hand.

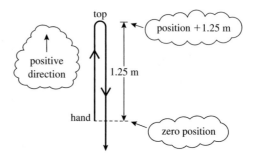

Figure 1.1

When it reaches the top, the marble might have travelled a distance of 1.25 m. Relative to your hand its position is then 1.25 m upwards or +1.25 m.

At the instant it returns to the same level as your hand it will have travelled a total distance of 2.5 m. Its *position*, however, is zero upwards.

A position is always referred to a fixed origin but a *displacement* can be measured from any position. When the marble returns to the level of your hand, its displacement is zero relative to your hand but –1.25 m relative to the top.

 What are the positions of the particles A, B and C in the diagram below?

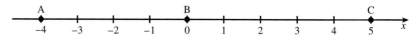

Figure 1.2

What is the displacement of B
(i) relative to A   (ii) relative to C?

Diagrams and graphs

In mathematics, it is important to use words precisely, even though they might be used more loosely in everyday life. In addition, a picture in the form of a diagram or graph can often be used to show the information more clearly.

Figure 1.3 is a *diagram* showing the direction of motion of the marble and relevant distances. The direction of motion is indicated by an arrow. Figure 1.4 is a *graph* showing the position above the level of your hand against the time. Notice that it is *not* the path of the marble.

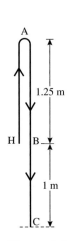

Figure 1.3

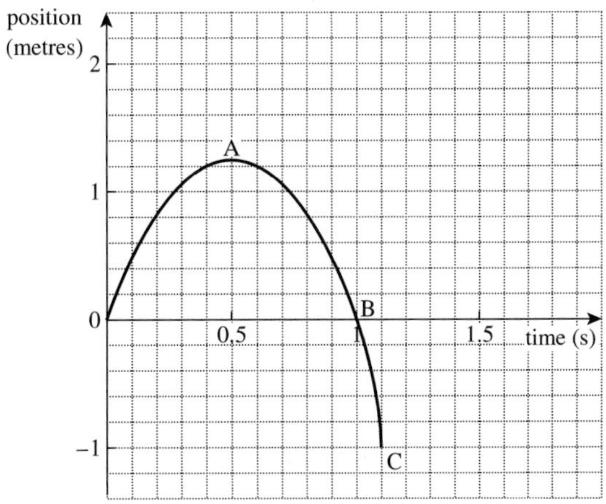

Figure 1.4

❓ The graph in figure 1.4 shows that the position is negative after one second (point B). What does this negative position mean?

Note

When drawing a graph it is very important to specify your axes carefully. Graphs showing motion usually have time along the horizontal axis. Then you have to decide where the origin is and which direction is positive on the vertical axis. In this graph the origin is at hand level and upwards is positive. The time is measured from the instant the marble leaves your hand.

Notation and units

As with most mathematics, you will see in this book that certain letters are commonly used to denote certain quantities. This makes things easier to follow. Here the letters used are:

- s, h, x, y and z for position
- t for time measured from a starting instant
- u and v for velocity
- a for acceleration.

The S.I. (Système International d'Unités) unit for *distance* is the metre (m), that for *time* is the second (s) and that for *mass* the kilogram (kg). Other units follow from these so speed is measured in metres per second, written ms^{-1}. S.I. units are used almost entirely in this book but occasional references are made to imperial and other units.

1 When the origin for the motion of the marble (see figure 1.3) is on the ground. What is its position
   (i) when it leaves your hand?   1
   (ii) at the top?   2.25

2 A boy throws a ball vertically upwards so that its position *y* m at time *t* is as shown in the graph.
   (i) Write down the position of the ball at times *t* = 0, 0.4, 0.8, 1.2, 1.6 and 2.   3.6 6 6.9 6 3.5 0
   (ii) Calculate the displacement of the ball relative to its starting position at these times.
   (iii) What is the total distance travelled
      (a) during the first 0.8 s   **(b)** during the 2 s of the motion?
         16.5   15

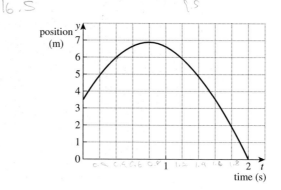

3 The position of a particle moving along a straight horizontal groove is given by *x* = 2 + *t*(*t* − 3) (0 ⩽ *t* ⩽ 5) where *x* is measured in metres and *t* in seconds.
   (i) What is the position of the particle at times *t* = 0, 1, 1.5, 2, 3, 4 and 5?
   (ii) Draw a diagram to show the path of the particle, marking its position at these times.
   (iii) Find the displacement of the particle relative to its initial position at *t* = 5.
   (iv) Calculate the total distance travelled during the motion.

4 For each of the following situations sketch a graph of position against time. Show clearly the origin and the positive direction.
   (i) A stone is dropped from a bridge which is 40 m above a river.
   (ii) A parachutist jumps from a helicopter which is hovering at 2000 m. She opens her parachute after 10 s of free fall.
   (iii) A bungee jumper on the end of an elastic string jumps from a high bridge.

5 The diagram is a sketch of the position–time graph for a fairground ride.

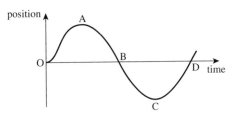

(i) Describe the motion, stating in particular what happens at O, A, B, C and D.

(ii) What type of ride is this?

Speed and velocity

Speed is a scalar quantity and does not involve direction. *Velocity* is the vector related to speed; its magnitude is the speed but it also has a direction. When an object is moving in the negative direction, its velocity is negative.

Amy has to post a letter on her way to college. The post box is 500 m east of her house and the college is 2.5 km to the west. Amy cycles at a steady speed of $10\,\text{ms}^{-1}$ and takes 10 s at the post box to find the letter and post it.

The diagram shows Amy's journey using east as the positive direction. The distance of 2.5 km has been changed to metres so that the units are consistent.

Figure 1.5

After she leaves the post box Amy is travelling west so her velocity is negative. It is $-10\,\text{ms}^{-1}$.

The distances and times for the three parts of Amy's journey are:

Home to post box 500 m $\frac{500}{10} = 50\,s$
At post box 0 m 10 s
Post box to college 3000 m $\frac{3000}{10} = 300\,s$

These can be used to draw the position–time graph using home as origin as in figure 1.6.

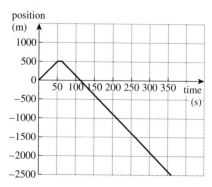

Figure 1.6

❓ Calculate the gradient of the three portions of this graph. What conclusions can you draw?

The velocity is the rate at which the position changes.

Velocity is represented by the gradient of the position–time graph.

Figure 1.7 is the velocity–time graph.

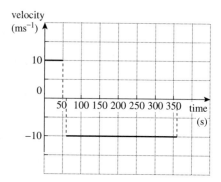

Note

By drawing the graphs below each other with the same horizontal scales, you can see how they correspond to each other.

Figure 1.7

Distance–time graphs

Figure 1.8 is the distance–time graph of Amy's journey. It differs from the position–time graph because it shows how far she travels irrespective of her direction. There are no negative values.

The gradient of this graph represents Amy's speed rather than her velocity.

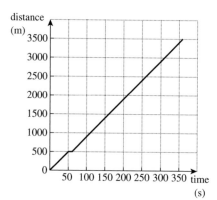

Figure 1.8

It has been assumed that Amy starts and stops instantaneously. What would more realistic graphs look like? Would it make a lot of difference to the answers if you tried to be more realistic?

Average speed and average velocity

You can find Amy's average speed on her way to college by using the definition

$$average\ speed = \frac{total\ distance\ travelled}{total\ time\ taken}$$

When the distance is in metres and the time in seconds, speed is found by dividing metres by seconds and is written as ms^{-1}. So Amy's average speed is

$$\frac{3500\,m}{360\,s} = 9.72\,ms^{-1}.$$

Amy's average velocity is different. Her displacement from start to finish is −2500 m so

The college is in the negative direction

$$average\ velocity = \frac{displacement}{time\ taken}$$

$$= \frac{-2500}{360} = -6.94\,ms^{-1}$$

If Amy had taken the same time to go straight from home to college at a steady speed, this steady speed would have been $6.94\,ms^{-1}$.

Velocity at an instant

The position–time graph for a marble thrown straight up into the air at 5 ms⁻¹ is curved because the velocity is continually changing.

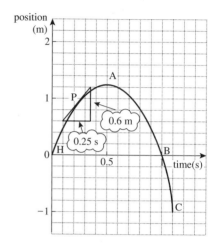

The velocity is represented by the gradient of the position–time graph. When a position–time graph is curved like this you can find *the velocity at an instant* of time by drawing a tangent as in figure 1.9. The velocity at P is approximately

$$\frac{0.6}{0.25} = 2.4 \,\text{ms}^{-1}$$

Figure 1.9

The velocity–time graph is shown in figure 1.10

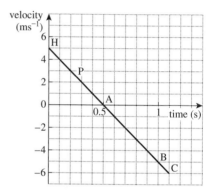

Figure 1.10

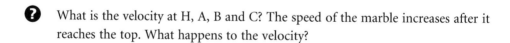 What is the velocity at H, A, B and C? The speed of the marble increases after it reaches the top. What happens to the velocity?

At the point A, the velocity and gradient of the position–time graph are zero. We say the marble is *instantaneously at rest*. The velocity at H is positive because the marble is moving in the positive direction (upwards). The velocity at B and at C is negative because the marble is moving in the negative direction (downwards).

EXERCISE 1B

1 Draw a speed–time graph for Amy's journey on page 5.

2 The distance–time graph shows the relationship between distance travelled and time for a person who leaves home at 9.00 am, walks to a bus stop and catches a bus into town.

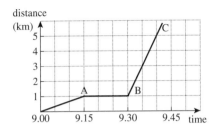

(i) Describe what is happening during the time from A to B.

(ii) The section BC is much steeper than OA; what does this tell us about the motion?

(iii) Draw the speed–time graph for the person.

(iv) What simplifications have been made in drawing these graphs?

3 For each of the following journeys find

(a) the initial and final positions;

(b) the total displacement;

(c) the total distance travelled;

(d) the velocity and speed for each part of the journey;

(e) the average velocity for the whole journey;

(f) the average speed for the whole journey.

(i)

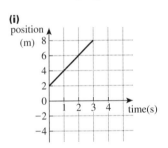

(ii)

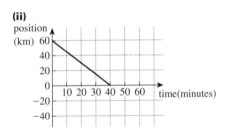

(iii)

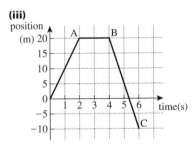

(iv)

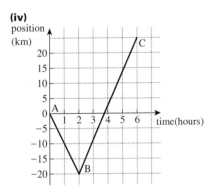

4 A plane flies from London to Toronto, a distance of 3560 miles at an average speed of 800 mph. It returns at an average speed of 750 mph. Find the average speed for the round trip.

Acceleration

In everyday language, the word 'accelerate' is usually used when an object speeds up and 'decelerate' when it slows down. The idea of deceleration is sometimes used in a similar way by mathematicians but in mathematics the word *acceleration* is used whenever there is a change in velocity, whether an object is speeding up, slowing down or changing direction. Acceleration is *the rate at which the velocity changes*.

Over a period of time

$$average\ acceleration = \frac{change\ in\ velocity}{time}$$

Acceleration is represented by the gradient of a velocity–time graph. It is a vector and can take different signs in a similar way to velocity. This is illustrated by Tom's cycle journey which is shown in Figure 1.11.

Tom turns on to the main road at $4\,\mathrm{ms^{-1}}$, accelerates uniformly, maintains a constant speed and then slows down uniformly to stop when he reaches home.

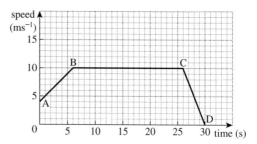

Figure 1.11

Between A and B, Tom's velocity increases by $(10 - 4) = 6\,\mathrm{ms^{-1}}$ in 6 seconds, that is by 1 metre per second every second.

This acceleration is written as $1\,\mathrm{ms^{-2}}$ (one metre per second squared) and is the gradient of AB.

From B to C   acceleration $= 0\,\mathrm{ms^{-2}}$ ◄—— (There is no change in velocity)

From C to D   acceleration $= \dfrac{(0 - 10)}{(30 - 26)} = -2.5\,\mathrm{ms^{-2}}$

From C to D, Tom is slowing down while still moving in the positive direction towards home, so his acceleration, the gradient of the graph, is negative.

The sign of acceleration

Think again about the marble thrown up into the air with a speed of $5\,\mathrm{ms^{-1}}$.

Figure 1.12 represents the velocity when *upwards* is taken as the positive direction and shows that the velocity *decreases* from $+5\,\mathrm{ms^{-1}}$ to $-5\,\mathrm{ms^{-1}}$ in 1 second.

This means that the gradient, and hence the acceleration, is *negative*. It is $-10\,\mathrm{ms^{-2}}$. (You might recognise the number 10 as an approximation to g. See Chapter 2 page 26.)

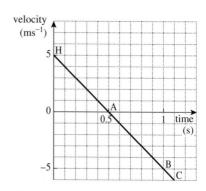

Figure 1.12

A car accelerates away from a set of traffic lights. It accelerates to a maximum speed and at that instant starts to slow down to stop at a second set of lights. Which of the graphs below could represent

(i) the distance–time graph
(ii) the velocity–time graph
(iii) the acceleration–time graph of its motion?

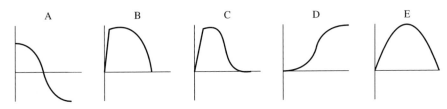

A   B   C   D   E

Figure 1.13

---

1 (i) Calculate the acceleration for each part of the following journey.
  (ii) Use your results to sketch an acceleration–time graph.

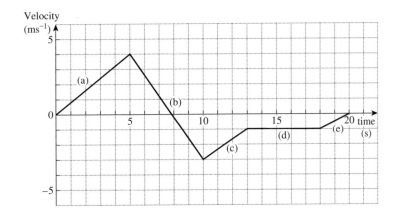

2 A particle moves so that its position x metres at time t seconds is
$x = 2t^3 - 18t$.
  (i) Calculate the position of the particle at times $t = 0, 1, 2, 3$ and 4.
  (ii) Draw a diagram showing the position of the particle at these times.
  (iii) Sketch a graph of the position against time.
  (iv) State the times when the particle is at the origin and describe the direction in which it is moving at those times.

3 A train on a Cornish branch line takes 45 minutes to complete its 15 mile trip. It stops for 1 minute at each of 7 stations during the trip.
  (i) Calculate the average speed of the train.
  (ii) What would be the average speed if the stop at each station was reduced to 20 seconds?

4 When Louise is planning car journeys she reckons that she can cover distances along main roads at roughly 60 mph and those in towns at 20 mph.
   (i) Find her average speed for each of the following journeys.
      (a) 20 miles to London and then 10 miles across town;
      (b) 150 miles to the coast and then 2 miles in town;
      (c) 20 miles to London and then 20 miles across town.
   (ii) In what circumstances would her average speed be 40 mph?

5 A lift travels up and down between the ground floor (G) and the roof garden (R) of a hotel. It starts from rest, takes 5 s to increase its speed uniformly to $2 \, ms^{-1}$, maintains this speed for 5 s and then slows down uniformly to rest in another 5 s. In the following questions, use upwards as positive.
   (i) Sketch a velocity–time graph for the journey from G to R.
   On one occasion the lift stops for 5 s at R before returning to G.
   (ii) Sketch a velocity–time graph for this journey from G to R and back.
   (iii) Calculate the acceleration for each 5 s interval. Take care with the signs.
   (iv) Sketch an acceleration–time graph for this journey.

6 A film of a dragster doing a 400 m run from a standing start yields the following positions at 1 second intervals.

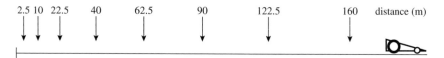

   (i) Draw a displacement–time graph of its motion.
   (ii) Use your graph to help you to sketch
      (a) the velocity–time graph
      (b) the acceleration–time graph.

Using areas to find distances and displacements

These distance–time and speed–time graphs model the motion of a stone falling from rest.

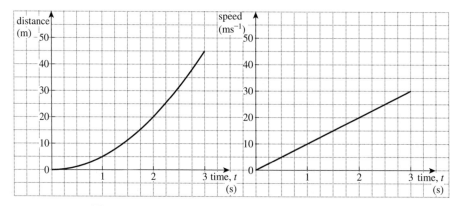

Figure 1.14   **Figure 1.15**

? Calculate the area between the speed–time graph and the x axis from
(i) $t = 0$ to 1   **(ii)** $t = 0$ to 2   **(iii)** $t = 0$ to 3.
Compare your answers with the distance that the stone has fallen, shown on the
distance–time graph, at $t = 1, 2$ and 3. What conclusions do you reach?

> *The area between a speed–time graph and the x axis represents the distance travelled.*

There is further evidence for this if you consider the units on the graphs.
Multiplying metres per second by seconds gives metres. A full justification relies
on the calculus methods you will learn in Chapter 8.

Finding the area under speed–time graphs

Many of these graphs consist of straight-line sections. The area is easily found by
splitting it up into triangles, rectangles or trapezia.

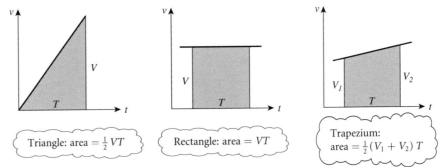

Triangle: area $= \frac{1}{2}VT$ Rectangle: area $= VT$ Trapezium: area $= \frac{1}{2}(V_1 + V_2)T$

Figure 1.16

EXAMPLE 1.1

The graph shows Tom's journey
from the time he turns on to the
main road until he arrives home.
How far does Tom cycle?

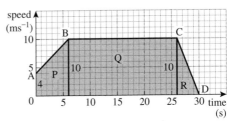

Figure 1.17

SOLUTION

The area under the speed–time graph
is found by splitting it into three regions.

P	trapezium:	area $= \frac{1}{2}(4 + 10) \times 6 =$	42 m
Q	rectangle:	area $= 10 \times 20$	= 200 m
R	triangle:	area $= \frac{1}{2} \times 10 \times 4$	= 20 m
		total area	= 262 m

Tom cycles 262 m.

? What is the meaning of the area between a velocity–time graph and the x axis?

The area between a velocity–time graph and the x axis

EXAMPLE 1.2

David walks east for 6 s at 2 ms⁻¹ then west for 2 s at 1 ms⁻¹. Draw
(i)   a diagram of the journey
(ii) the speed–time graph
(iii) the velocity–time graph.

Interpret the area under each graph.

SOLUTION

(i)   David's journey is illustrated below.

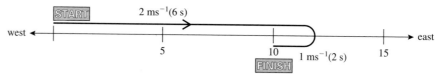

Figure 1.18

(ii) Speed–time graph

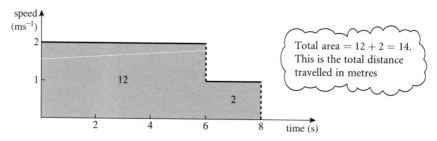

Total area = 12 + 2 = 14.
This is the total distance
travelled in metres

Figure 1.19

(iii) Velocity–time graph

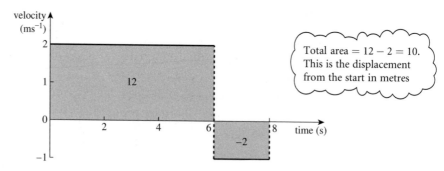

Total area = 12 − 2 = 10.
This is the displacement
from the start in metres

Figure 1.20

> *The area between a velocity–time graph and the x axis represents the change in position, that is the displacement.*

When the velocity is negative, the area is below the axis and represents a
displacement in the negative direction, west in this case.

Estimating areas

Sometimes the velocity–time graph does not consist of straight lines so you have to make the best estimate you can be counting the squares underneath it or by replacing the curve by a number of straight lines as for the trapezium rule (see *Pure Mathematics 1*).

? This speed–time graph shows the motion of a dog over a 60 s period.

Estimate how far the dog travelled during this time.

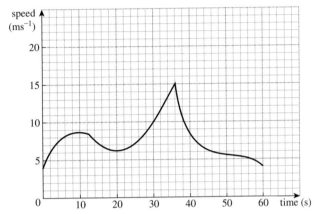

Figure 1.21

EXAMPLE 1.3

On the London underground, Oxford Circus and Piccadilly Circus are 0.8 km apart. A train accelerates uniformly to a maximum speed when leaving Oxford Circus and maintains this speed for 90 s before decelerating uniformly to stop at Piccadilly Circus. The whole journey takes 2 minutes. Find the maximum speed.

SOLUTION

The sketch of the speed–time graph of the journey shows all the information available with suitable units. The maximum speed is $v\,\text{ms}^{-1}$.

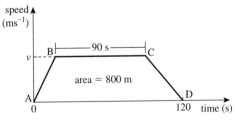

Figure 1.22

The area is $\frac{1}{2}(120 + 90) \times v = 800$

$$v = \frac{800}{105}$$

$$= 7.619$$

The maximum speed of the train is $7.6\,\text{ms}^{-1}$.

? Does it matter how long the train takes to speed up and slow down?

1 The graphs show the speeds of two cars travelling along a street.

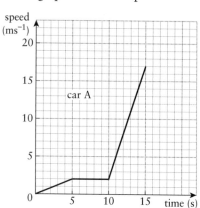

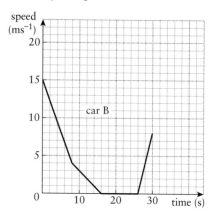

For each car find

(i) the acceleration for each part of its motion

(ii) the total distance it travels in the given time

(iii) its average speed.

2 The graph shows the speed of a lorry when it enters a very busy motorway.

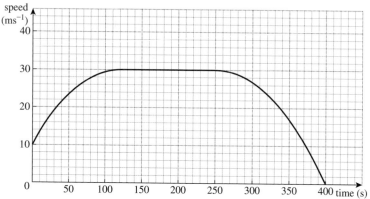

(i) Describe the journey over this time.

(ii) Use a ruler to make a tangent to the graph and hence estimate the acceleration at the beginning and end of the period.

(iii) Estimate the distance travelled and the average speed.

3 A train leaves a station where it has been at rest and picks up speed at a constant rate for 60 s. It then remains at a constant speed of 17 ms^{-1} for 60 s before it begins to slow down uniformly as it approaches a set of signals. After 45 s it is travelling at 10 ms^{-1} and the signal changes. The train again increases speed uniformly for 75 s until it reaches a speed of 20 ms^{-1}. A second set of signals then orders the train to stop, which it does after slowing down uniformly for 30 s.

(i) Draw a speed–time graph for the train.

(ii) Use your graph to find the distance that it has travelled from the station.

4   When a parachutist jumps from a helicopter hovering above an airfield her speed increases at a constant rate to 28 ms^{-1} in the first 3 s of her fall. It then decreases uniformly to 8 ms^{-1} in a further 6 s, remaining constant until she reaches the ground.

(i)   Sketch a speed–time graph for the parachutist.

(ii) Find the height of the plane when the parachutist jumps out if the complete jump takes 1 minute.

5   A car is moving at 20 ms^{-1} when it begins to increase speed. Every 10 s it gains 5 ms^{-1} until it reaches its maximum speed of 50 ms^{-1} which it retains.

(i)   Draw the speed–time graph of the car.

(ii)   When does the car reach its maximum speed of 50 ms^{-1}?

(iii) Find the distance travelled by the car after 150 s.

(iv)   Write down expressions for the speed of the car t seconds after it begins to speed up.

6   A train takes 10 minutes to travel from Birmingham New Street to Birmingham International. The train accelerates at a rate of 0.5 ms^{-2} for 30 s. It then travels at a constant speed and is finally brought to rest in 15 s with a constant deceleration.

(i)   Sketch a velocity–time graph for the journey.

(ii) Find the steady speed, the rate of deceleration and the distance from Birmingham New Street to Birmingham International.

7   A train was scheduled to travel at 50 ms^{-1} for 15 minutes on part of its journey. The velocity–time graph illustrates the actual progress of the train which was forced to stop because of signals.

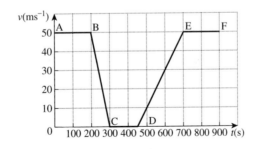

(i)   Without carrying out any calculations, describe what was happening to the train in each of the stages BC, CD and DE.

(ii)   Find the deceleration of the train while it was slowing down and the distance travelled during this stage.

(iii) Find the acceleration of the train when it starts off again and the distance travelled during this stage.

(iv)   Calculate by how long the stop will have delayed the train.

(v)   Sketch the distance–time graph for the journey between A and F, marking the points A, B, C, D, E and F.   [MEI]

8 A car is travelling at 36 km h^{-1} when the driver has to perform an emergency stop. During the time the driver takes to appreciate the situation and apply the brakes the car has travelled 7 m ('thinking distance'). It then pulls up with constant deceleration in a further 8 m ('braking distance') giving a total stopping distance of 15 m.

(i) Find the initial speed of the car in metres per second and the time that the driver takes to react.

(ii) Sketch the velocity–time graph for the car.

(iii) Calculate the deceleration once the car starts braking.

(iv) What is the stopping distance for a car travelling at 60 km h^{-1} if the reaction time and the deceleration are the same as before?

INVESTIGATION

TRAIN JOURNEY

If you look out of a train window you will see distance markers beside the track every quarter of a mile. Take a train journey and record the time as you go past each marker. Use your figures to draw distance–time, speed–time and acceleration–time graphs. What can you conclude about the greatest acceleration, deceleration and speed of the train?

KEY POINTS

1 Vectors (with magnitude and direction)   **Scalars** (magnitude only)

Vectors	Scalars
Displacement	Distance
Position – displacement from a fixed origin	
Velocity – rate of change of position	Speed – magnitude of velocity
Acceleration – rate of change of velocity	
	Time

- *Vertical* is towards the centre of the earth; *horizontal* is perpendicular to vertical.

2 Diagrams
- Motion along a line can be illustrated vertically or horizontally (as shown)

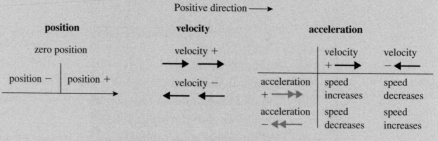

- *Average speed* $= \dfrac{\text{total distance travelled}}{\text{total time taken}}$

- *Average velocity* $= \dfrac{\text{displacement}}{\text{time taken}}$

3 Graphs

- *Position–time*

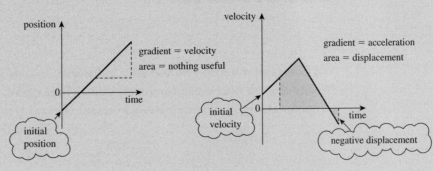

- *Velocity–time*

gradient = acceleration
area = displacement

initial velocity

negative displacement

- *Distance–time*

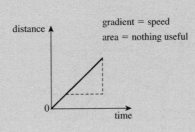

gradient = speed
area = nothing useful

- *Speed–time*

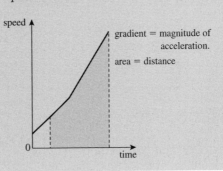

gradient = magnitude of acceleration.
area = distance

2

Modelling using constant acceleration

The poetry of motion! The real way to travel! The only way to travel!
Here today – in next week tomorrow! Villages skipped, towns and cities
jumped – always somebody else's horizon! O bliss! O poop-poop! O my!

Kenneth Grahame (Wind in the Willows)

Setting up a mathematical model

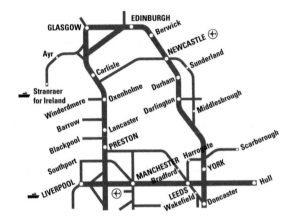

Figure 2.1 *Northern Great Britain's Main Railways*

? The figure shows part of the map of the main railway lines of northern Great
Britain. Which of the following statements can you be sure of just by looking at
this map? Which of them might be important to a visitor from abroad?

(i) Darlington is on the line from York to Durham.
(ii) Carlisle is nearer to Glasgow than it is to Newcastle.
(iii) Leeds is due east of Manchester.
(iv) The quickest way from Leeds to Glasgow is via York.

This is a *diagrammatic model* of the railway system which gives essential though
by no means all the information you need for planning train journeys. You can
be sure about the places a line passes through but distances and directions are
only approximate and if you compare this map with an ordinary map you will
see that statements (ii) and (iii) are false. You will also need further information
from timetables to plan the best way to get from Leeds to Glasgow.

Making simplifying assumptions

When setting up a model, you first need to decide what is essential. For example, what would you take into account and what would you ignore when considering the motion of a car travelling from Bristol to London?

You will need to know the distance and the time taken for parts of the journey, but you might decide to ignore the dimensions of the car and the motion of the wheels. You would then be using the idea of a *particle* to model the car. *A particle has no dimensions.*

You might also decide to ignore the bends in the road and its width and so treat it as a *straight line with only one dimension*. A length along the line would represent a length along the road in the same way as a piece of thread following a road on a map might be straightened out to measure its length.

You might decide to split the journey up into parts and assume that the speed is constant over these parts.

The process of making decisions like these is called *making simplifying assumptions* and is the first stage of setting up a *mathematical model* of the situation.

Defining the variables and setting up the equations

The next step in setting up a mathematical model is to *define the variables* with suitable units. These will depend on the problem you are trying to solve. Suppose you want to know where you ought to be at certain times in order to maintain a good average speed between Bristol and London. You might define your variables as follows:

- the total time since the car left Bristol is t hours
- the distance from Bristol at time t is x km
- the average speed up to time t is v km h^{-1}.

Then, at Newbury $t = t_1$ and $x = x_1$; etc.

You can then *set up equations* and go through the mathematics required to solve the problem. Remember to check that your answer is sensible. If it isn't, you might have made a mistake in your mathematics or your simplifying assumptions might need reconsideration.

The theories of mechanics that you will learn about in this course, and indeed any other studies in which mathematics is applied, are based on mathematical models of the real world. When necessary, these models can become more complex as your knowledge increases.

❓ The simplest form of the Bristol to London model assumes that the speed remains constant over sections of the journey. Is this reasonable?

For a much shorter journey, you might need to take into account changes in the velocity of the car. This chapter develops the mathematics required when an object can be modelled as a *particle moving in a straight line with constant acceleration*. In most real situations this is only the case for part of the motion – you wouldn't expect a car to continue accelerating at the same rate for very long – but it is a very useful model to use as a first approximation over a short time.

The constant acceleration formulae

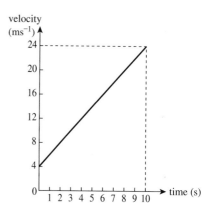

The velocity–time graph shows part of the motion of a car on a fairground ride as it picks up speed. The graph is a straight line so the velocity increases at a constant rate and the car has a constant acceleration which is equal to the gradient of the graph.

The velocity increases from $4\,\mathrm{ms}^{-1}$ to $24\,\mathrm{ms}^{-1}$ in $10\,\mathrm{s}$ so its acceleration is

$$\frac{24 - 4}{10} = 2\,\mathrm{ms}^{-2}.$$

Figure 2.2

In general, when the initial velocity is $u\,\mathrm{ms}^{-1}$ and the velocity a time $t\,\mathrm{s}$ later is $v\,\mathrm{ms}^{-1}$, as in figure 2.3, the increase in velocity is $(v - u)\,\mathrm{ms}^{-1}$ and the constant acceleration $a\,\mathrm{ms}^{-2}$ is given by

$$\frac{v - u}{t} = a$$

so $\qquad v - u = at$

$$v = u + at. \qquad ①$$

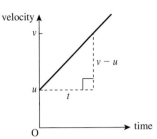

Figure 2.3

The area under the graph represents the distance travelled. For the fairground car, that is represented by a trapezium of area

$$\frac{(4 + 24)}{2} \times 10 = 140 \text{ m}.$$

In the general situation, the area represents the displacement s metres and is

$$s = \frac{(u + v)}{2} \times t \qquad ②$$

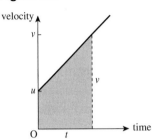

Figure 2.4

? The two equations, ① and ②, can be used as formulae for solving problems when the acceleration is constant. Check that they work for the fairground ride.

There are other useful formulae as well. For example, you might want to find the displacement, s, without involving v in your calculations. This can be done by looking at the area under the velocity–time graph in a different way, using the rectangle R and the triangle T. In the diagram

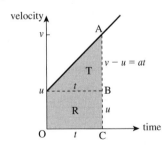

Figure 2.5

$$AC = v \text{ and } BC = u$$

so $\qquad AB = v - u$

$$= at \qquad \text{from equation } ①$$

$$\text{total area} = \text{area of R} + \text{area of T}$$

so $\qquad s = ut + \frac{1}{2} \times t \times at$

Giving $\qquad s = ut + \frac{1}{2}at^2 \qquad ③$

To find a formula which does not involve t, you need to eliminate t. One way to do this is first to rewrite equations ① and ② as

$$v - u = at \quad \text{and} \quad v + u = \frac{2s}{t}$$

and then multiplying them gives

$$(v - u)(v + u) = at \times \frac{2s}{t}$$
$$v^2 - u^2 = 2as$$
$$\boxed{v^2 = u^2 + 2as} \hspace{3cm} ④$$

You might have seen the equations ① to ④ before. They are sometimes called the *suvat* equations or formulae and they can be used whenever an object can be assumed to be moving with *constant acceleration.*

 When solving problems it is important to remember the requirement for constant acceleration and also to remember to specify positive and negative directions clearly.

EXAMPLE 2.1

A bus leaving a bus stop accelerates at $0.8 \, \text{ms}^{-2}$ for 5 s and then travels at a constant speed for 2 minutes before slowing down uniformly at $4 \, \text{ms}^{-2}$ to come to rest at the next bus stop. Calculate

(i) the constant speed
(ii) the distance travelled while the bus is accelerating
(iii) the total distance travelled.

SOLUTION

(i) The diagram shows the information for the first part of the motion.

| | | | 0.8 ms⁻² | |
acceleration 0.8 ms^{-2}

velocity 0 ms^{-1} v ms^{-1}

time   A   5 s   B

distance s_1m

Figure 2.6

Let the constant speed be $v \, \text{ms}^{-1}$.

$u = 0$, $a = 0.8$, $t = 5$, so use $v = u + at$

$$v = 0 + 0.8 \times 5$$
$$= 4$$

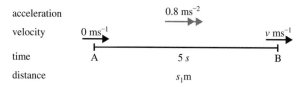

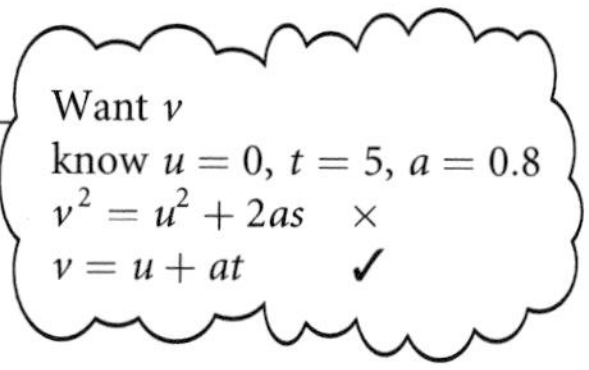

Want v
know $u = 0$, $t = 5$, $a = 0.8$
$v^2 = u^2 + 2as$ ✗
$v = u + at$ ✓

The constant speed is $4 \, \text{ms}^{-1}$.

(ii) Let the distance travelled be s_1 m.

> Use the suffix because there are three distances to be found in this question

$u = 0$, $a = 0.8$, $t = 5$, so use $s = ut + \frac{1}{2}at^2$

> Want s
> know $u = 0$, $t = 5$, $a = 0.8$
> $s = \frac{1}{2}(u+v)t$ ✗
> $s = ut + \frac{1}{2}at^2$ ✓

$$s_1 = 0 + \frac{1}{2} \times 0.8 \times 5^2$$
$$= 10$$

The bus accelerates over 10 m.

(iii) The diagram gives all the information for the rest of the journey.

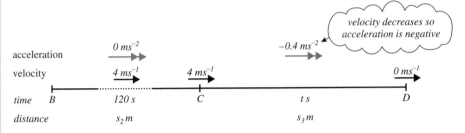

> velocity decreases so acceleration is negative

Figure 2.7

Between B and C the velocity is constant so the distance travelled is $4 \times 120 = 480$ m.

Let the distance between C and D be s_3 m.

$u = 4$, $a = -0.4$, $v = 0$, so use $v^2 = u^2 + 2as$

> Want s
> know $u = 4$, $a = -0.4$, $v = 0$
> $v = u + at$ ✗
> $s = ut + \frac{1}{2}at^2$ ✗
> $s = \frac{1}{2}(u+v)t$ ✗
> $v^2 = u^2 + 2as$ ✓

$$0 = 16 + 2(-0.4)s_3$$
$$0.8s_3 = 16$$
$$s_3 = 20$$

Distance taken to slow down $= 20$ m

The total distance travelled is
$(10 + 480 + 20)$ m $= 510$ m.

Units in the *suvat* equations

Constant acceleration usually takes place over short periods of time so it is best to use ms^{-2} for this. When you don't need to use a value for the acceleration you can, if you wish, use the *suvat* equations with other units provided they are consistent. This is shown in the next example.

EXAMPLE 2.2

When leaving a town, a car accelerates from 30 mph to 60 mph in 5 s. Assuming the acceleration is constant, find the distance travelled in this time.

SOLUTION

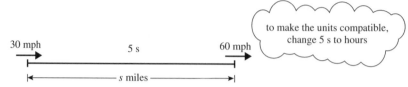

to make the units compatible, change 5 s to hours

Figure 2.8

Let the distance travelled be s miles. We want s and are given $u = 30$, $v = 60$ and $t = 5 \div 3600$ so we need an equation involving u, v, t and s.

$$s = \frac{(u + v)}{2} \times t$$

$$s = \frac{(30 + 60)}{2} \times \frac{5}{3600}$$

$$= \tfrac{1}{16}$$

The distance travelled is $\frac{1}{16}$ mile or 110 yards. (One mile is 1760 yards.)

❓ Write all the measurements in example 2.2 in terms of metres and seconds and then find the resulting distance. (110 yards is about 100 m.)

⚠ In the examples 2.1 and 2.2, the bus and the car are always travelling in the positive direction so it is safe to use s for distance. Remember that s is not the same as the distance travelled if the direction changes during the motion.

The acceleration due to gravity

When a model ignoring air resistance is used, all objects falling freely under gravity fall with the same constant acceleration, g ms^{-2}. This varies over the surface of the earth. In this book it is assumed that all the situations occur in a place where it is 9.8 ms^{-2} or sometimes 10 ms^{-2} as an approximation. Most answers are given correct to three significant figures so that you can check your working.

EXAMPLE 2.3

A coin is dropped from rest at the top of a building of height 12 m and travels in a straight line with constant acceleration $10\,\text{ms}^{-2}$.
Find the time it takes to reach the ground and the speed of impact.

SOLUTION

Suppose the time taken to reach the ground is t seconds. Using SI units, $u = 0$, $a = 10$ and $s = 12$ when the coin hits the ground, so we need to use a formula involving u, a, s and t.

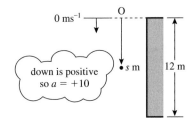

Figure 2.9

$$s = ut + \tfrac{1}{2}at^2$$
$$12 = 0 + \tfrac{1}{2} \times 10 \times t^2$$
$$t^2 = 2.4$$
$$t = 1.55$$

To find the velocity, v, a formula involving s, u, a and v is required.

$$v^2 = u^2 + 2as$$
$$v^2 = 0 + 2 \times 10 \times 12$$
$$v^2 = 240$$
$$v = 15.5$$

The coin takes 1.55 s to hit the ground and has speed $15.5\,\text{ms}^{-1}$ on impact.

Summary

The equations for motion with *constant acceleration* are

① $\quad v = u + at$ ② $\quad s = \dfrac{(u + v)}{2} \times t$

③ $\quad s = ut + \tfrac{1}{2}at^2$ ④ $\quad v^2 = u^2 + 2as$

? Derive equation ③ algebraically by substituting for v from equation ① into equation ②.

If you look at these equations you will see that each omits one variable. But there are five variables and only four equations; there isn't one without u. An equation omitting u is

⑤ $\quad s = vt - \tfrac{1}{2}at^2$

? How can you derive this by referring to a graph or using substitution?

When using these equations make sure that the units you use are consistent. For example, when the time is t seconds and the distance s metres, any speed involved is in ms^{-1}.

1 **(i)** Find v when $u = 10$, $a = 6$ and $t = 2$.
   (ii) Find s when $v = 20$, $u = 4$ and $t = 10$.
   (iii) Find s when $v = 10$, $a = 2$ and $t = 10$.
   (iv) Find a when $v = 2$, $u = 12$, $s = 7$.

2 Decide which equation to use in each of these situations.

   (i) Given u, s, a; find v.   **(ii)** Given a, u, t; find v.
   (iii) Given u, a, t; find s.   **(iv)** Given u, v, s; find t.
   (v) Given u, s, v; find a.   **(vi)** Given u, s, t; find a.
   (vii) Given u, a, v; find s.   **(viii)** Given a, s, t; find v.

3 Assuming no air resistance, a ball has an acceleration of $9.8\,ms^{-2}$ when it is dropped from a window (so its initial speed, when $t = 0$, is zero). Calculate:
   (i) its speed after $1\,s$ and after $10\,s$.
   (ii) how far it has fallen after $1\,s$ and after $10\,s$.
   (iii) how long it takes to fall $19.6\,m$.

   Which of these answers are likely to need adjusting to take account of air resistance? Would you expect your answer to be an over- or underestimate?

4 A car starting from rest at traffic lights reaches a speed of $90\,km\,h^{-1}$ in $12\,s$. Find the acceleration of the car (in ms^{-2}) and the distance travelled. Write down any assumptions that you have made.

5 A top sprinter accelerates from rest to $9\,ms^{-1}$ in $2\,s$. Calculate his acceleration, assumed constant, during this period and the distance travelled.

6 A van skids to a halt from an initial speed of $24\,ms^{-1}$ covering a distance of $36\,m$. Find the acceleration of the van (assumed constant) and the time it takes to stop.

7 An object moves along a straight line with acceleration $-8\,ms^{-2}$. It starts its motion at the origin with velocity $16\,ms^{-1}$.
   (i) Write down equations for its position and velocity at time $t\,s$.
   (ii) Find the smallest non-zero time when
      (a) the velocity is zero
      (b) the object is at the origin.
   (iii) Sketch the position–time, velocity–time and speed–time graphs for
      $0 \leqslant t \leqslant 4$.

Further examples

The next two examples illustrate ways of dealing with more complex problems. In example 2.4, none of the possible equations has only one unknown and there are also two situations, so simultaneous equations are used.

EXAMPLE 2.4

James practises using the stopwatch facility on his new watch by measuring the time between lamp posts on a car journey. As the car speeds up, two consecutive times are 1.2 s and 1 s. Later he finds out that the lamp posts are 30 m apart.

(i) Calculate the acceleration of the car (assumed constant) and its speed at the first lamp post.

(ii) Assuming the same acceleration, find the time the car took to travel the 30 m before the first lamp post.

SOLUTION

(i) The diagram shows all the information assuming the acceleration is $a\,\text{ms}^{-2}$ and the velocity at A is $u\,\text{ms}^{-1}$.

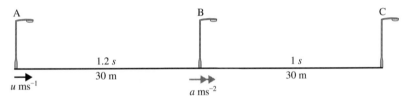

Figure 2.10

For AB, $s = 30$ and $t = 1.2$. We are using u and we want a so we use

$$s = ut + \tfrac{1}{2}at^2$$
$$30 = 1.2u + \tfrac{1}{2}a \times 1.2^2$$
$$30 = 1.2u + 0.72a \qquad \text{①}$$

To use the same equation for the part BC we would need the velocity at B and this brings in another unknown. It is much better to go back to the beginning and consider the whole of AC with $s = 60$ and $t = 2.2$. Then again using $s = ut + \tfrac{1}{2}at^2$

$$60 = 2.2u + \tfrac{1}{2}a \times 2.2^2$$
$$60 = 2.2u + 2.42a \qquad \text{②}$$

These two simultaneous equations in two unknowns can be solved more easily if they are simplified. First make the coefficients of u integers.

	①×10 ÷ 12	$25 = u + 0.6a$	③
	②×5	$300 = 11u + 12.1a$	④
then	③×11	$275 = 11u + 6.6a$	⑤

Subtracting gives

$$25 = 0 + 5.5a$$
$$a = 4.545$$

Now substitute 4.545 for a in ③ to find

$$u = 25 - 0.6 \times 4.545 = 22.273.$$

The acceleration of the car is 4.55 ms^{-2} and the initial speed is 22.3 ms^{-1} (correct to 3 sf).

(ii)

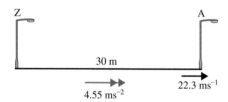

Figure 2.11

For this part, we know that $s = 30$, $v = 22.3$ and $a = 4.55$ and we want t so we use the fifth formula.

$$s = vt - \tfrac{1}{2}at^2$$
$$30 = 22.3 \times t - \tfrac{1}{2} \times 4.55 \times t^2$$
$$\Rightarrow \qquad 2.275t^2 - 22.3t + 30 = 0$$

Solving this using the quadratic formula gives $t = 1.61$ and $t = 8.19$.

The most sensible answer to this particular problem is 1.61 s.

❓ Calculate u when $t = 8.19$, $v = 22.3$ and $a = 4.55$. Is $t = 8.19$ a possible answer?

Using a non-zero initial displacement

What, in the constant acceleration equations, are v and s when $t = 0$?

Putting $t = 0$ in the *suvat* equations gives the *initial values*, u for the velocity and $s = 0$ for the position.

Sometimes, however, it is convenient to use an origin which gives a non-zero value for s when $t = 0$. For example, when you model the motion of a rubber thrown vertically upwards you might decide to find its height above the ground rather than from the point from which it was thrown.

What is the effect on the various *suvat* equations if the initial position is s_0 rather than 0?

If the height of the rubber above the ground is s at time t and s_0 when $t = 0$, the displacement over time t is $s - s_0$. You then need to replace equation ③ with

$$s - s_0 = ut + \tfrac{1}{2}at^2$$

The next example avoids this in the first part but it is very useful in part (ii).

EXAMPLE 2.5

A juggler throws a ball up in the air with initial speed $5\,\text{ms}^{-1}$ from a height of $1.2\,\text{m}$. It has a constant acceleration of $10\,\text{ms}^{-2}$ vertically downwards due to gravity.

(i) Find the maximum height of the ball above the ground and the time it takes to reach it.

At the instant that the ball reaches its maximum height, the juggler throws up another ball with the same speed and from the same height.

(ii) Where and when will the balls pass each other?

SOLUTION

(i) In this example it is very important to draw a diagram and to be clear about the position of the origin. When O is 1.2 m above the ground and s is the height in metres above O after t s, the diagram looks like figure 2.12.

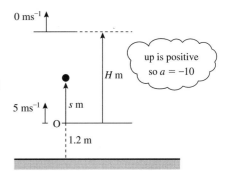

up is positive so $a = -10$

Figure 2.12

At the point of maximum height, let $s = H$ and $t = t_1$. ◄

Use the suffix because there are two times to be found in this question.

The ball stops instantaneously before falling so at the top $v = 0$.

An equation involving u, v, a and s is required.

$$v^2 = u^2 + 2as$$
$$0 = 5^2 + 2 \times (-10) \times H$$
$$H = 1.25$$

The acceleration given is constant, $a = -10$; $u = +5$; $v = 0$ and $s = H$

The maximum height of the ball above the ground is $1.25 + 1.2 = 2.45\,\text{m}$.

To find t_1, given $v = 0$, $a = -10$ and $u = +5$ requires a formula in v, u, a and t.

$$v = u + at$$
$$0 = 5 + (-10)\,t_1$$
$$t_1 = 0.5$$

The ball takes half a second to reach its maximum height.

(ii) Now consider the motion from the instant the first ball reaches the top of its path and the second is thrown up.

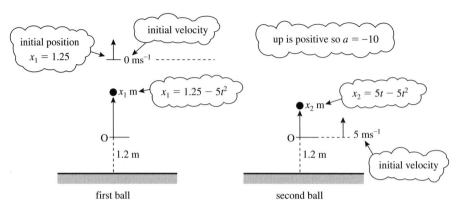

Figure 2.13

Suppose that the balls have displacements *above* the origin of x_1 m and x_2 m, as shown in the diagram, at a general time t s after the second ball is thrown up. The initial position of the second ball is 0, but the initial position of the first ball is $+1.25$ m.

For each ball we know u and a. We want to involve t and s so we use
$$s - s_0 = ut + \tfrac{1}{2}at^2$$

i.e. $$s = s_0 + ut + \tfrac{1}{2}at^2$$

For the first ball:
$$x_1 = 1.25 + 0 \times t + \tfrac{1}{2} \times (-10) \times t^2$$

This makes $x_1 = 1.25$ when $t = 0$

$$x_1 = 1.25 - 5t^2 \qquad ①$$

x_1 decreases as t increases

For the second ball:
$$x_2 = 0 + 5 \times t + \tfrac{1}{2} \times (-10) \times t^2$$
$$x_2 = 5t - 5t^2 \qquad ②$$

Suppose the balls pass after a time t s. This is when they are at the same height, so equate x_1 and x_2 from equations ① and ②.

$$1.25 - 5t^2 = 5t - 5t^2$$
$$1.25 = 5t$$
$$t = 0.25$$

Then substituting $t = 0.25$ in ① and ② gives

$$x_1 = 1.25 - 5 \times 0.25^2 = 0.9375$$

These are the same, as expected

and

$$x_2 = 5 \times 0.25 - 5 \times 0.25^2 = 0.9375$$

The balls pass after 0.25 seconds at a height of $1.2 + 0.94$ m $= 2.14$ m above the ground (correct to the nearest centimetre).

? Try solving part (ii) of this example by supposing that the first ball falls x m and the second rises $1.25 - x$ m in t seconds.

> *Note*
>
> The balls pass after half the time to reach the top, but *not* half way up.

? Why don't they travel half the distance in half the time?

Use $g = 9.8\,ms^{-2}$ *in this exercise unless otherwise specified.*

1 A car is travelling along a straight road. It accelerates uniformly from rest to a speed of $15\,\text{ms}^{-1}$ and maintains this speed for 10 minutes. It then decelerates uniformly to rest. If the acceleration and deceleration are $5\,\text{ms}^{-2}$ and $8\,\text{ms}^{-2}$ respectively, find the total journey time and the total distance travelled during the journey.

2 A skier pushes off at the top of a slope with an initial speed of $2\,\text{ms}^{-1}$. She gains speed at a constant rate throughout her run. After $10\,\text{s}$ she is moving at $6\,\text{ms}^{-1}$.
   (i) Find an expression for her speed t seconds after she pushes off.
   (ii) Find an expression for the distance she has travelled at time t seconds.
   (iii) The length of the ski slope is $400\,\text{m}$. What is her speed at the bottom of the slope?

3 Towards the end of a half-marathon Sabina is $100\,\text{m}$ from the finish line and is running at a constant speed of $5\,\text{ms}^{-1}$. Daniel, who is $140\,\text{m}$ from the finish and is running at $4\,\text{ms}^{-1}$, decides to accelerate to try to beat Sabina. If he accelerates uniformly at $0.25\,\text{ms}^{-2}$ does he succeed?

4 Rupal throws a ball upwards at $2\,\text{ms}^{-1}$ from a window which is $4\,\text{m}$ above ground level.
   (i) Write down an equation for the height h m of the ball above the ground after t s (while it is still in the air).
   (ii) Use your answer to part (i) to find the time the ball hits the ground.
   (iii) How fast is the ball moving just before it hits the ground?
   (iv) In what way would you expect your answers to parts (ii) and (iii) to change if you were able to take air resistance into account?

5 Nathan hits a tennis ball straight up into the air from a height of 1.25 m above the ground. The ball hits the ground after 2.5 seconds. Assuming $g = 10 \, \text{ms}^{-2}$, find

(i) the speed Nathan hits the ball

(ii) the greatest height above the ground reached by the ball

(iii) the speed the ball hits the ground

(iv) how high the ball bounces if it loses 0.2 of its speed on hitting the ground.

(v) Is your answer to part (i) likely to be an over- or under-estimate given that you have ignored air resistance?

6 A ball is dropped from a building of height 30 m and at the same instant a stone is thrown vertically upwards from the ground so that it hits the ball. In modelling the motion of the ball and stone it is assumed that each object moves in a straight line with a constant downward acceleration of magnitude $10 \, \text{ms}^{-2}$. The stone is thrown with initial speed of $15 \, \text{ms}^{-1}$ and is h_s metres above the ground t seconds later.

(i) Draw a diagram of the ball and stone before they collide, marking their positions.

(ii) Write down an expression for h_s at time t.

(iii) Write down an expression for the height h_b of the ball at time t.

(iv) When do the ball and stone collide?

(v) How high above the ground do the ball and stone collide?

7 When Kim rows her boat, the two oars are both in the water for 3 s and then both out of the water for 2 s. This 5 s cycle is then repeated. When the oars are in the water the boat accelerates at a constant $1.8 \, \text{ms}^{-2}$ and when they are not in the water it decelerates at a constant $2.2 \, \text{ms}^{-2}$.

(i) Find the change in speed that takes place in each 3 s period of acceleration.

(ii) Find the change in speed that takes place in each 2 s period of deceleration.

(iii) Calculate the change in the boat's speed for each 5 s cycle.

(iv) A race takes Kim 45 s to complete. If she starts from rest what is her speed as she crosses the finishing line?

(v) Discuss whether this is a realistic speed for a rowing boat.

8 A ball is dropped from a tall building and falls with acceleration of magnitude $10 \, \text{ms}^{-2}$. The distance between floors in the block is constant. The ball takes 0.5 s to fall from the 14th to the 13th floor and 0.3 s to fall from the 13th floor to the 12th. What is the distance between floors?

9 Two clay pigeons are launched vertically upwards from exactly the same spot at 1 s intervals. Each clay pigeon has initial speed $30 \, \text{ms}^{-1}$ and acceleration $10 \, \text{ms}^{-2}$ downwards. How high above the ground do they collide?

10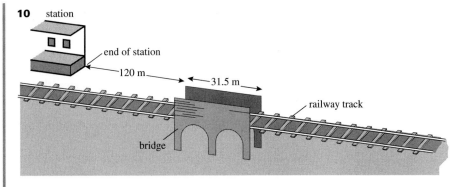

A train accelerates along a straight, horizontal section of track. The driver notes that he reaches a bridge 120 m from the station in 8 s and that he crosses the bridge, which is 31.5 m long, in a further 2 s.

The motion of the train is modelled by assuming constant acceleration. Take the speed of the train when leaving the station to be u ms^{-1} and the acceleration to have the value a ms^{-2}.

(i) By considering the part of the journey from the station to the bridge, show that $u + 4a = 15$.

(ii) Find a second equation involving u and a.

(iii) Solve the two equations for u and a to show that a is 0.15 and find the value of u.

(iv) If the driver also notes that he travels 167 m in the 10 s after he crosses the bridge, have you any evidence to reject the modelling assumption that the acceleration is constant?

[**MEI**]

INVESTIGATION

The situation described below involves mathematical modelling. You will need to take these steps to help you.

(i) Make a list of the assumptions you need to make to simplify the situation to the point where you can apply mathematics to it.

(ii) Make a list of the quantities involved.

(iii) Find out any information you require such as safe stopping distances or a value for the acceleration and deceleration of a car on a housing estate.

(iv) Assign suitable letters for your unknown quantities. (Don't vary too many things at once.)

(v) Set up your equations and solve them. You might find it useful to work out several values and draw a suitable graph.

(vi) Decide whether your results make sense, preferably by checking them against some real data.

(vii) If you think your results need adjusting, decide whether any of your initial assumptions should be changed and, if so, in what way.

SPEED BUMPS

The residents of a housing estate are worried about the danger from cars being driven at high speed. They request that speed bumps be installed.

How far apart should the bumps be placed to ensure that drivers do not exceed a speed of 30 mph? Some of the things to consider are the maximum sensible velocity over each bump and the time taken to speed up and slow down.

KEY POINTS

1 **The *suvat* equations**
   - The equations for motion with constant acceleration are

   ① $\quad v = u + at$ ② $\quad s = \dfrac{(u + v)}{2} \times t$

   ③ $\quad s = ut + \frac{1}{2}at^2$ ④ $\quad v^2 = u^2 + 2as$

   ⑤ $\quad s = vt - \frac{1}{2}at^2$

   - a is the constant acceleration; s is the displacement from the starting position at time t; v is the velocity at time t; u is the velocity when $t = 0$.

   If $s = s_0$ when $t = 0$, replace s in each equation with $(s - s_0)$.

2 **Vertical motion under gravity**
   - The acceleration due to gravity $(g\,\text{ms}^{-2})$ is $9.8\,\text{ms}^{-2}$ vertically downwards.
   - Always draw a diagram and decide in advance where your origin is and which way is positive.
   - Make sure that your units are compatible.

3 **Using a mathematical model**
   - Make simplifying assumptions by deciding what is most relevant.
     For example: a car is a *particle* with no dimensions
     a road is a *straight line* with one dimension
     acceleration is constant.
   - Define variables and set up equations.
   - Solve the equations.
   - Check that the answer is sensible. If not, think again.

3 Forces and Newton's laws of motion

Nature and Nature's Laws lay hid in Night.
God said, Let Newton be! and All was Light.

Alexander Pope

Force diagrams

The picture shows a crate of medical supplies being dropped into a remote area by parachute. What forces are acting on the crate of supplies and the parachute?

One force which acts on every object near the earth's surface is its own *weight*. This is the force of gravity pulling it towards the centre of the earth. The weight of the crate acts on the crate and the weight of the parachute acts on the parachute.

The parachute is designed to make use of *air resistance*. A resistance force is present whenever a solid object moves through a liquid or gas. It acts in the opposite direction to the motion and depends on the speed of the object. The crate also experiences air resistance, but to a lesser extent than the parachute.

Other forces are the *tensions* in the guy lines attaching the crate to the parachute. These pull upwards on the crate and downwards on the parachute.

All these forces can be shown most clearly if you draw *force diagrams* for the crate and the parachute.

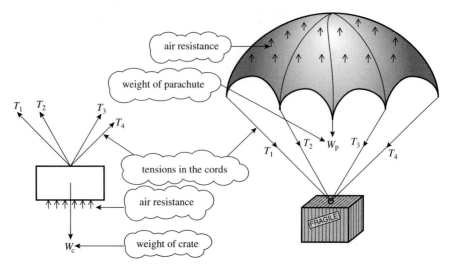

Figure 3.1 *Forces acting on the crate*   **Figure 3.2** *Forces acting on the parachute*

Force diagrams are essential for the understanding of most mechanical situations. A force is a vector: it has a magnitude, or size, and a direction. It also has a *line of action*. This line often passes through a point of particular interest. Any force diagram should show clearly

- the direction of the force
- a label showing the magnitude
- the line of action.

In figures 3.1 and 3.2 each force is shown by an arrow along its line of action. The air resistance has been depicted a lot of separate arrows but this is not very satisfactory. It is much better if the combined effect can be shown by one arrow. When you have learned more about vectors, you will see how the tensions in the guy lines can also be combined into one force if you wish. The forces on the crate and parachute can then be simplified.

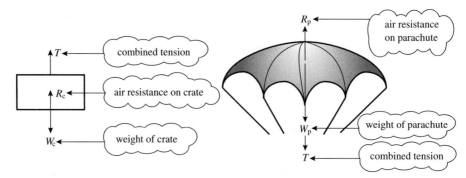

Figure 3.3 *Forces acting on the crate*   **Figure 3.4** *Forces acting on the parachute*

Centre of mass and the particle model

When you combine forces you are finding their *resultant*. The weights of the crate and parachute are also found by combining forces; they are the resultant of the weights of all their separate parts. Each weight acts through a point called the *centre of mass* or centre of gravity.

Think about balancing a pen on your finger. The diagrams show the forces acting on the pen.

Figure 3.5   **Figure 3.6**

So long as you place your finger under the centre of mass of the pen, as in figure 3.5, it will balance. There is a force called a *reaction* between your finger and the pen which balances the weight of the pen. The forces on the pen are then said to be *in equilibrium*. If you place your finger under another point, as in figure 3.6, the pen will fall. The pen can only be in equilibrium if the two forces have the same line of action.

If you balance the pen on two fingers, there is a reaction between each finger and the pen at the point where it touches the pen. These reactions can be combined into one resultant vertical reaction acting through the centre of mass.

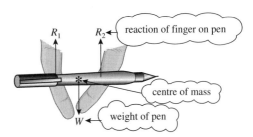

Figure 3.7

The behaviour of objects which are liable to rotate under the action of forces is covered in *Mechanics 2*. This book deals with situations where the resultant of the forces does not cause rotation. An object can then be modelled as a particle, that is a point mass, situated at its centre of mass.

Newton's third law of motion

Sir Isaac Newton (1642–1727) is famous for his work on gravity and the mechanics you learn in this course is often called Newtonian Mechanics because it is based entirely on Newton's three laws of motion. These laws provide us with an extremely powerful model of how objects, ranging in size from specks of dust to planets and stars, behave when they are influenced by forces.

We start with Newton's *third law* which says that

> *When one object exerts a force on another there is always a reaction of the same kind which is equal, and opposite in direction, to the acting force.*

You might have noticed that the combined tensions acting on the parachute and the crate in figures 3.3 and 3.4 are both marked with the same letter, *T*. The crate applies a force on the parachute through the supporting guy lines and the parachute applies an equal and opposite force on the crate. When you apply a force to a chair by sitting on it, it responds with an equal and opposite force on you. Figure 3.8 shows the forces acting when someone sits on a chair.

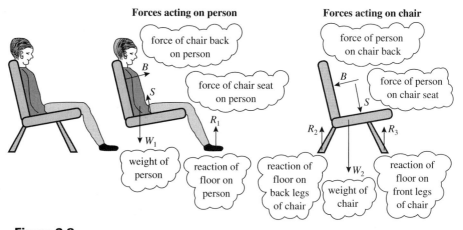

Forces acting on person

force of chair back on person

force of chair seat on person

B

S

R_1

W_1

weight of person

reaction of floor on person

Forces acting on chair

force of person on chair back

force of person on chair seat

B

S

R_2

R_3

reaction of floor on back legs of chair

W_2

weight of chair

reaction of floor on front legs of chair

Figure 3.8

The reactions of the floor on the chair and on your feet act where there is contact with the floor. You can use R_1, R_2 and R_3 to show that they have different magnitudes. There are equal and opposite forces acting on the floor, but the forces on the floor are not being considered so do not appear here.

 Why is the weight of the person *not* shown on the force diagram for the chair?

Gravitational forces obey Newton's third law just as other forces between bodies. According to Newton's universal law of gravitation, the earth pulls us towards its centre and we pull the earth in the opposite direction. However, in this book we are only concerned with the gravitational force on us and not the force we exert on the earth.

All the forces you meet in mechanics apart from the gravitational force are the result of physical contact. This might be between two solids or between a solid and a liquid or gas.

Friction and normal reaction

When you push your hand along a table, the table reacts in two ways.

- Firstly there are forces which stop your hand going through the table. Such forces are always present when there is any contact between your hand and the table. They are at right angles to the surface of the table and their resultant is called the *normal reaction* between your hand and the table.
- There is also another force which tends to prevent your hand from sliding. This is the *friction* and it acts in a direction which opposes the sliding.

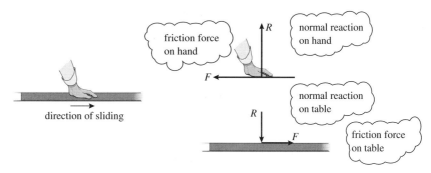

Figure 3.9

The diagram shows the reaction forces acting on your hand and on the table. By Newton's third law they are equal and opposite to each other. The frictional force is due to tiny bumps on the two surfaces (*see* electronmicrograph below). When you hold your hands together you will feel the normal reaction between them. When you slide them against each other you will feel the friction.

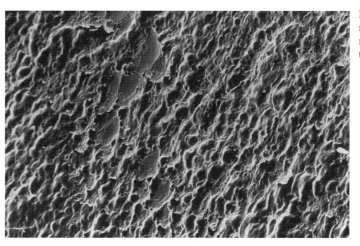

Etched glass magnified to high resolution, showing the tiny bumps.

When the friction between two surfaces is negligible, at least one of the surfaces is said to be *smooth*. This is a modelling assumption which you will meet frequently in this book. Oil can make surfaces smooth and ice is often modelled as a smooth surface.

When the contact between two surfaces is smooth, the only force between them is at right angles to any possible sliding and is just the normal reaction.

? What direction is the reaction between the sweeper's broom and the smooth ice?

EXAMPLE 3.1

A TV set is standing on a small table. Draw a diagram to show the forces acting on the TV and on the table as seen from the front.

SOLUTION

The diagram shows the forces acting on the TV and on the table. They are all vertical because the weights are vertical and there are no horizontal forces acting.

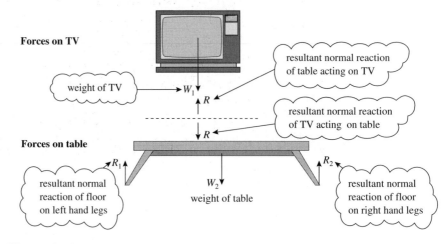

Figure 3.10

EXAMPLE 3.2

Draw diagrams to show the forces acting on a tennis ball which is hit downwards across the court:

(i) at the instant it is hit by the racket
(ii) as it crosses the net
(iii) at the instant it lands on the other side.

SOLUTION

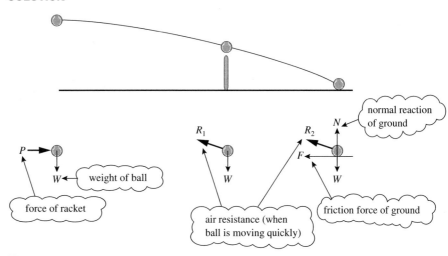

Figure 3.11

EXERCISE 3A

In this exercise draw clear diagrams to show the forces acting on the objects named in italics. Clarity is more important than realism when drawing these diagrams.

1 A *gymnast* hanging at rest on a bar.

2 A *light bulb* hanging from a ceiling.

3 A *book* lying at rest on a table.

4 A *book* at rest on a table but being pushed by a small horizontal force.

5 *Two books* lying on a table, one on top of the other.

6 A *horizontal plank* being used to bridge a stream.

7 A *snooker ball* on a table which can be assumed to be smooth
   (i) as it lies at rest on the table.
   (ii) at the instant it is hit by the cue.

8 An *ice hockey puck*
   (i) at the instant it is hit when standing on smooth ice.
   (ii) at the instant it is hit when standing on rough ice.

9 A *cricket ball* which follows the path shown on the right.
Draw diagrams for each of the three positions A, B and C (include air resistance).

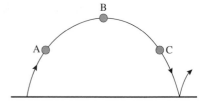

10 (i) *Two balls* colliding in mid-air.
(ii) *Two balls* colliding on a snooker table.

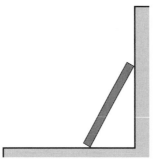

11 A *paving stone* leaning against a wall

12 A *cylinder* at rest on smooth surfaces.

Force and motion

? How are the rails and handles provided in buses and trains used by standing passengers?

Newton's first law

Newton's *first law* can be stated as follows:

> *Every particle continues in a state of rest or uniform motion in a straight line unless acted on by a resultant external force.*

Newton's first law provides a reason for the handles on trains and buses. When you are on a train which is stationary or moving at constant speed in a straight line you can easily stand without support. But when the velocity of the train changes a force is required to change your velocity to match. This happens when the train slows down or speeds up. It also happens when the train goes round a bend even if the speed does not change. The velocity changes because the direction changes.

? Why is Josh's car in the pond?

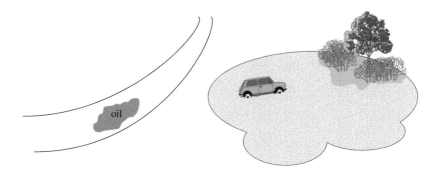

oil

Figure 3.12

EXAMPLE 3.3

A £1 coin is balanced on your finger and then you move it upwards.

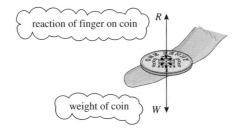

reaction of finger on coin

weight of coin

Figure 3.13

By considering Newton's first law, what can you say about *W* and *R* in these situations?

(i) The coin is stationary.

(ii) The coin is moving upwards with a constant velocity.

(iii) The speed of the coin is increasing as it moves upwards.

(iv) The speed of the coin is decreasing as it moves upwards.

SOLUTION

(i) When the coin is stationary the velocity does not change. The forces are in equilibrium and $R = W$.

(ii) When the coin is moving upwards with a constant velocity the velocity does not change. The forces are in equilibrium and $R = W$.

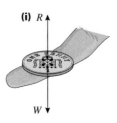

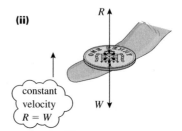

Figure 3.14   **Figure 3.15**

(iii) When the speed of the coin is increasing as it moves upwards there must be a net upwards force to make the velocity increase in the upwards direction so $R > W$. The net force is $R - W$.

(iv) When the speed of the coin is decreasing as it moves upwards there must be a net downwards force to make the velocity decrease and slow the coin down as it moves upwards. In this case $W > R$ and the net force is $W - R$.

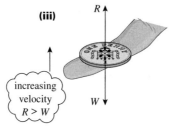

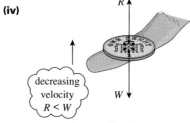

Figure 3.16   **Figure 3.17**

EXERCISE 3B

1 A book is resting on an otherwise empty table.
   (i) Draw diagrams showing the forces acting on
      (a) the book
      (b) the table as seen from the side.
   (ii) Write down equations connecting the forces acting on the book and on the table.

2 You balance a coin on your finger and move it up and down. The reaction of your finger on the coin is R and its weight is W. Decide in each case whether R is greater than, less than or equal to W and describe the net force.
   (i) The coin is moving downwards with a constant velocity.
   (ii) The speed of the coin is increasing as it moves downwards.
   (iii) The speed of the coin is decreasing as it moves downwards.

3 In each of the following situations say whether the forces acting on the object are in equilibrium by deciding whether its motion is changing.

(i)   A car that has been stationary, as it moves away from a set of traffic lights.

(ii) A motorbike as it travels at a steady 60 mph along a straight road.

(iii) A parachutist descending at a constant rate.

(iv) A box in the back of a lorry as the lorry picks up speed along a straight, level motorway.

(v)   An ice hockey puck sliding across a smooth ice rink.

(vi) A book resting on a table.

(vii) A plane flying at a constant speed in a straight line, but losing height at a constant rate.

(viii) A car going round a corner at constant speed.

4 Explain each of the following in terms of Newton's laws.

(i)   Seat belts must be worn in cars.

(ii) Head rests are necessary in a car to prevent neck injuries when there is a collision from the rear.

Driving forces and resistances to the motion of vehicles

In problems about such things as cycles, cars and trains, all the forces acting along the line of motion will usually be reduced to two or three, the *driving force* forwards, the *resistance* to motion (air resistance, etc.) and possibly a *braking force* backwards.

Resistances due to air or water always act in a direction opposite to the velocity of a vehicle or boat and are usually more significant for fast-moving objects.

Tension and thrust

The lines joining the crate of supplies to the parachute described at the beginning of this chapter are in tension. They pull upwards on the crate and downwards on the parachute. You are familiar with tensions in ropes and strings, but rigid objects can also be in tension.

When you hold the ends of a pencil, one with each hand, and pull your hands apart, you are pulling on the pencil. What is the pencil doing to each of your hands? Draw the forces acting on your hands and on the pencil.

Now draw the forces acting on your hands and on the pencil when you push the pencil inwards.

Your first diagram might look like figure 3.18. The pencil is in tension so there is an inward *tension* force on each hand.

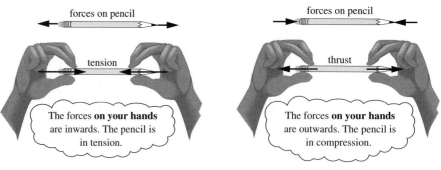

forces on pencil

forces on pencil

tension

thrust

The forces **on your hands** are inwards. The pencil is in tension.

The forces **on your hands** are outwards. The pencil is in compression.

Figure 3.18

Figure 3.19

When you push the pencil inwards the forces on your hands are outwards as in figure 3.19. The pencil is said to be *in compression* and the outward force on each hand is called a *thrust*.

If each hand applies a force of 2 units on the pencil, the tension or thrust acting on each hand is also 2 units because each hand is in equilibrium.

 Which of the above diagrams is still possible if the pencil is replaced by a piece of string?

Resultant forces and equilibrium

You have already met the idea that a single force can have the same effect as several forces acting together. Imagine that several people are pushing a car. A single rope pulled by another car can have the same effect. The force of the rope is equivalent to the resultant of the forces of the people pushing the car. When there is no resultant force, the forces are in equilibrium and there is no change in motion.

EXAMPLE 3.4

A car is using a tow bar to pull a trailer along a straight, level road. There are resisting forces R acting on the car and S acting on the trailer. The driving force of the car is D and its braking force is B.

Draw diagrams showing the horizontal forces acting on the car and the trailer

(i) when the car is moving at constant speed
(ii) when the speed of the car is increasing
(iii) when the car brakes and slows down rapidly.

In each case write down the resultant force acting on the car and on the trailer.

SOLUTION

(i) When the car moves at constant speed, the forces are as shown in figure 3.20. The tow bar is in tension and the effect is a forward force on the trailer and an equal and opposite backwards force on the car.

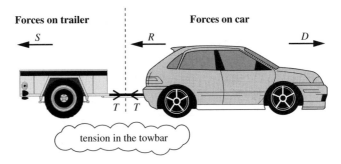

Figure 3.20 *Car travelling at constant speed*

There is no resultant force on either the car or the trailer when the speed is constant; the forces on each are in equilibrium.

For the trailer: $T - S = 0$

For the car: $D - R - T = 0$

(ii) When the car speeds up, the same diagram will do, but now the magnitudes of the forces are different. There is a resultant *forwards* force on both the car and the trailer.

For the trailer: resultant $= T - S$

For the car: resultant $= D - R - T$

(iii) When the car brakes a resultant *backwards* force is required to slow down the trailer. When the resistance S is not sufficiently large to do this, a thrust in the tow bar comes into play as shown in the figure 3.21.

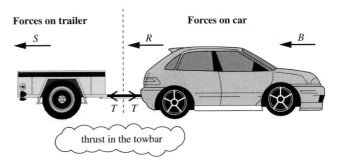

Figure 3.21 *Car braking*

For the trailer: resultant $= T + S$

For the car: resultant $= B + R - T$

Newton's second law

Newton's *second law* gives us more information about the relationship between the magnitude of the resultant force and the change in motion. Newton said that

> *The change in motion is proportional to the force.*

For objects with constant mass, this can be interpreted as *the force is proportional to the acceleration.*

Resultant force = a constant × acceleration   ①

The constant in this equation is proportional to the mass of the object, a more massive object needs a larger force to produce the same acceleration. For example, you and your friends would be able to give a car a greater acceleration than a lorry.

Newton's second law is so important that a special unit of force, the *newton (N)*, has been defined so that the constant in equation ① is actually equal to the mass. A force of 1 newton will give a mass of 1 kilogram an acceleration of $1 \, \text{ms}^{-2}$. The equation then becomes:

$$\text{Resultant force} = \text{mass} \times \text{acceleration} \qquad ②$$

This is written: $F = ma$

The resultant force and the acceleration are always in the same direction.

Relating mass and weight

The *mass* of an object is related to the amount of matter in the object. It is a *scalar*. The *weight* of an object is a force. It has magnitude and direction and so is a *vector*.

The mass of an astronaut on the moon is the same as his mass on the earth but his weight is only about one-sixth of his weight on the earth. This is why he can bounce around more easily on the moon. The gravitational force on the moon is less because the mass of the moon is less than that of the earth.

When Buzz Aldrin made the first landing on the moon in 1969 with Neil Armstrong, one of the first things he did was to drop a feather and a hammer to demonstrate that they fell at the same rate. Their accelerations due to the gravitational force of the moon were equal, even though they had very different masses. The same is true on earth. If other forces were negligible all objects would fall with an acceleration g.

When the weight is the only force acting on an object, Newton's second law means that

Weight in newtons = mass in kg $\times$ g in ms^{-2}.

Using standard letters: $\quad\boxed{W = mg}$

Even when there are other forces acting, the weight can still be written as mg. A good way to visualise a force of 1 N is to think of the weight of an apple. 1 kg of apples weighs (1×9.8) N that is approximately 10 N. There are about 10 small to medium-sized apples in 1 kg, so each apple weighs about 1 N.

> *Note*
>
> Anyone who says 1 kg of apples *weighs* 1 kg is not strictly correct. The terms weight and mass are often confused in everyday language but it is very important for your study of mechanics that you should understand the difference.

EXAMPLE 3.5

What is the weight of

(i) a baby of mass 3 kg?
(ii) a golf ball, mass 46 g?

SOLUTION

(i) The baby's weight is $3 \times 9.8 = 29.4$ N
(ii) Mass of golf ball $= 46$ g
$\qquad\qquad\qquad = 0.046$ kg
Weight $\qquad\qquad = 0.046 \times 9.8$ N
$\qquad\qquad\qquad = 0.45$ N (to 2 sf)

Data: On the earth g $= 9.8\,ms^{-2}$. *On the moon* g $= 1.6\,ms^{-2}$.
1000 newtons (N) $= 1$ *kilonewton* (kN).

1 Calculate the magnitude of the force of gravity on the following objects on
  the earth.
  (i) A suitcase of mass 15 kg.
  (ii) A car of mass 1.2 tonnes. (1 tonne $= 1000$ kg)
  (iii) A letter of mass 50 g.

2 Find the mass of each of these objects on the earth.
  (i) A girl of weight 600 N.
  (ii) A lorry of weight 11 kN.

3 A person has mass 65 kg. Calculate the force of gravity
  (i) of the earth on the person
  (ii) of the person on the earth.

4 What reaction force would an astronaut of mass 70 kg experience while
  standing on the moon?

5 Two balls of the same shape and size but with masses 1 kg and 3 kg are
  dropped from the same height.
  (i) Which hits the ground first?
  (ii) If they were dropped on the moon what difference would there be?

6 **(i)** Estimate your mass in kilograms.
  (ii) Calculate your weight when you are on the earth's surface.
  (iii) What would your weight be if you were on the moon?
  (iv) When people say that a baby weighs 4 kg, what do they mean?

? Most weighing machines have springs or some other means to measure force
even though they are calibrated to show mass. Would something appear to
weigh the same on the moon if you used one of these machines? What could
you use to find the mass of an object irrespective of where you measure it?

Pulleys

In the remainder of this chapter weight will be represented by *mg*. You will learn
to apply Newton's second law more generally in the next chapter.

A pulley can be used to change the direction of a force; for example it is much
easier to pull down on a rope than to lift a heavy weight. When a pulley is well
designed it takes a relatively small force to make it turn and such a pulley is
modelled as being *smooth and light*. Whatever the direction of the string passing
over this pulley, its tension is the same on both sides.

Figure 3.22 shows the forces acting when a pulley is used to lift a heavy parcel.

Forces acting on the ends of the rope

Forces acting on the pulley

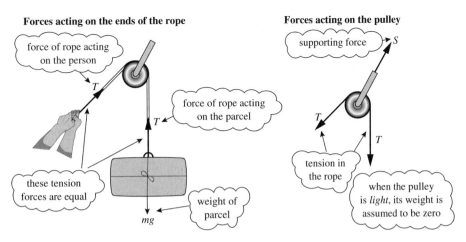

force of rope acting
on the person

supporting force

S

T

force of rope acting
on the parcel

T

T

T

these tension
forces are equal

tension in
the rope

when the pulley
is *light*, its weight is
assumed to be zero

weight of
parcel

mg

Figure 3.22

Note

The rope is in tension. It is not possible for a rope to exert a thrust force.

EXAMPLE 3.6

In this diagram the pulley is smooth and light
and the 2 kg block, A, is on a rough surface.

(i) Draw diagrams to show the forces acting on
each of A and B.

(ii) If the block A does not slip, find the tension
in the string and calculate the magnitude of the
friction force on the block.

(iii) Write down the resultant force acting on each
of A and B if the block slips and accelerates.

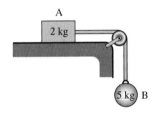

A

2 kg

5 kg B

Figure 3.23

SOLUTION

(i)

Forces on A

Forces on pulley

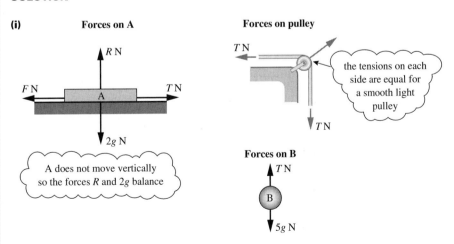

R N

F N

A

T N

$2g$ N

A does not move vertically
so the forces R and $2g$ balance

T N

the tensions on each
side are equal for
a smooth light
pulley

T N

Forces on B

T N

B

$5g$ N

Figure 3.24

Note

The masses of 2 kg and 5 kg are not shown in the force diagram. The weights $2g$ N and $5g$ N are more appropriate.

(ii) When the block does not slip, the forces on B are in equilibrium so
$$5g - T = 0$$
$$T = 5g$$

The tension throughout the string is $5g$ N.

For A, the resultant horizontal force is zero so
$$T - F = 0$$
$$F = T = 5g$$

The friction force is $5g$ N towards the left.

(iii) When the block slips, the forces are not in equilibrium and T and F have different magnitudes.

The resultant horizontal force on A is $(T - F)$ N towards the right.
The resultant force on B is $(5g - T)$ N vertically downwards.

EXERCISE 3D

In this exercise you are asked to draw force diagrams using the various types of force you have met in this chapter. Remember that all the forces you need, other than weight, occur when objects are in contact or joined together in some way. Where motion is involved, indicate its direction clearly.

1 Draw labelled diagrams showing the forces acting on the objects in *italics*.
   (i) *A car* towing a caravan.
   (ii) *A caravan* being towed by a car.
   (iii) *A person* pushing a supermarket trolley.
   (iv) *A suitcase* on a horizontal moving pavement (as at an airport)
        (a) immediately after it has been put down
        (b) when it is moving at the same speed as the pavement.
   (v) *A sledge* being pulled uphill.

2 Ten boxes each of mass 5 kg are stacked on top of each other on the floor.
   (i) What forces act on the top box?
   (ii) What forces act on the bottom box?

3 The diagrams show a box of mass m under different systems of forces.
   (i) In the first case the box is at rest. State the value of F_1
   (ii) In the second case the box is slipping. Write down the resultant horizontal force acting on it.

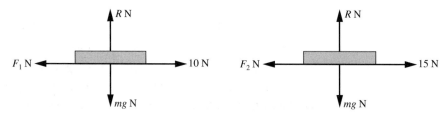

4 In this diagram the pulleys are smooth and light, the strings are light, and the table is rough.

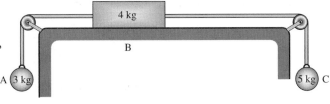

(i) What is the direction of the friction force on the block B?

(ii) Draw clear diagrams to show the forces on each of A, B and C.

(iii) By considering the equilibrium of A and C, calculate the tensions in the strings when there is no slipping.

(iv) Calculate the magnitude of the friction when there is no slipping.

Now suppose that there is insufficient friction to stop the block from slipping.

(v) Write down the resultant force acting on each of A, B and C.

5 A man who weighs 720 N is doing some repairs to a shed. In each of these situations draw diagrams showing

(a) the forces the man exerts on the shed

(b) all the forces acting on the man (ignore any tools he might be using).

In each case, compare the reaction between the man and the floor with his weight of 720 N.

(i) He is pushing upwards on the ceiling with force U N.

(ii) He is pulling downwards on the ceiling with force D N.

(iii) He is pulling upwards on a nail in the floor with force F N.

(iv) He is pushing downwards on the floor with force T N.

6 The diagram shows a train, consisting of an engine of mass 50 000 kg pulling two trucks, A and B, each of mass 10 000 kg. The force of resistance on the engine is 2000 N and that on each of the trucks 200 N. The train is travelling at constant speed.

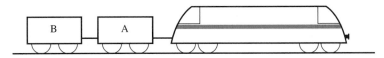

(i) Draw a diagram showing the horizontal forces on the train as a whole. Hence, by considering the equilibrium of the train as a whole, find the driving force provided by the engine.

The coupling connecting truck A to the engine exerts a force T_1 N on the engine and the coupling connecting truck B to truck A exerts a force T_2 N on truck B.

(ii) Draw diagrams showing the horizontal forces on the engine and on truck B.

(iii) By considering the equilibrium of the engine alone, find T_1.

(iv) By considering the equilibrium of truck B alone, find T_2.

(v) Show that the forces on truck A are also in equilibrium.

Historical note

Isaac Newton was born in Lincolnshire in 1642. He was not an outstanding scholar either as a schoolboy or as a university student, yet later in life he made remarkable contributions in dynamics, optics, astronomy, chemistry, music theory and theology. He became Member of Parliament for Cambridge University and later Warden of the Royal Mint. His tomb in Westminster Abbey reads 'Let mortals rejoice That there existed such and so great an Ornament to the Human Race'.

KEY POINTS

1 **Newton's laws of motion**

   I Every object continues in a state of rest or uniform motion in a straight line unless it is acted on by a resultant external force.

   II Resultant force = mass × acceleration or $\mathbf{F} = m\mathbf{a}$

   III When one object exerts a force on another there is always a reaction which is equal, and opposite in direction, to the acting force.

   - *Force* is a vector; *mass* is a scalar.
   - The *weight* of an object is the force of gravity pulling it towards the centre of the earth. Weight $= mg$ vertically downwards.

2 **S.I. units**
   - length: metre (m)
   - time: second (s)
   - velocity: ms^{-1}
   - acceleration: ms^{-2}
   - mass: kilogram (kg)

3 **Force**
   1 newton (N) is the force required to give a mass of 1 kg an acceleration of $1\,\text{ms}^{-2}$.
   A force of 1000 newtons (N) = 1 kilonewton (kN).

4 Types of force

- Forces due to contact between surfaces

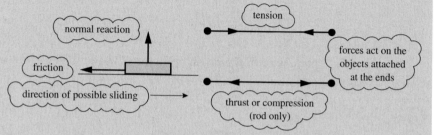

- Forces in a joining rod or string

- A smooth light pulley

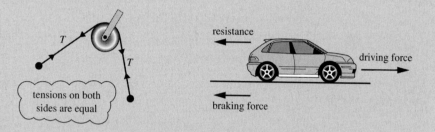

- Forces on a wheeled vehicle

5 Commonly used modelling terms

- inextensible   does not vary in length
- light   negligible mass
- negligible   small enough to ignore
- particle   negligible dimensions
- smooth   negligible friction
- uniform   the same throughout

4 Applying Newton's second law along a line

Nature to him was an open book. He stands before us, strong, certain and alone.

Einstein on Newton

Newton's second law

? Attach a weight to a spring balance and move it up and down. What happens to the pointer on the balance?

What would you observe if you stood on some bathroom scales in a moving lift?

Hold a heavy book on your hand and move it up and down.
What force do you feel on your hand?

Equation of motion

Suppose you make the book accelerate upwards at $a\,\text{ms}^{-2}$. Figure 4.1 shows the forces acting on the book and the acceleration.

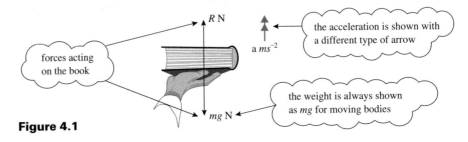

forces acting on the book

R N

$a\,ms^{-2}$

the acceleration is shown with a different type of arrow

the weight is always shown as mg for moving bodies

mg N

Figure 4.1

By Newton's first law, a resultant force is required to produce an acceleration. In this case the resultant upwards force is $R - mg$ newtons.

You were introduced to Newton's second law in Chapter 3. When the forces are in newtons, the mass in kilograms and the acceleration in metres per second squared, this law is:

$$\text{Resultant force} = \text{mass} \times a$$

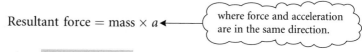

where force and acceleration are in the same direction.

So for the book: $R - mg = ma$ ①

When Newton's second law is applied, the resulting equation is called *the equation of motion*.

When you give a book of mass 0.8 kg an acceleration of 0.5 ms⁻² equation ① becomes

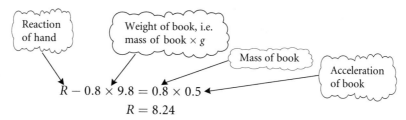

Reaction of hand

Weight of book, i.e. mass of book × g

Mass of book

Acceleration of book

$$R - 0.8 \times 9.8 = 0.8 \times 0.5$$
$$R = 8.24$$

When the book is accelerating upwards the reaction force of your hand on the book is 8.24 N. This is equal and opposite to the force experienced by you so the book feels heavier than its actual weight, mg, which is $0.8 \times 9.8 = 7.84$ N.

1 Calculate the resultant force in newtons required to produce the following accelerations.

(i)   A car of mass 400 kg has acceleration $2\,\text{ms}^{-2}$.

(ii) A blue whale of mass 177 tonnes has acceleration $\frac{1}{2}\,\text{ms}^{-2}$.

(iii) A pygmy mouse of mass 7.5 g has acceleration of $3\,\text{ms}^{-2}$.

(iv) A freight train of mass 42 000 tonnes brakes with deceleration of $0.02\,\text{ms}^{-2}$.

(v)   A bacterium of mass 2×10^{-16} g has acceleration $0.4\,\text{ms}^{-2}$.

(vi) A woman of mass 56 kg falling off a high building has acceleration $9.8\,\text{ms}^{-2}$.

(vii) A jumping flea of mass 0.05 mg accelerates at $1750\,\text{ms}^{-2}$ during take-off.

(viii) A galaxy of mass 10^{42} kg has acceleration $10^{-12}\,\text{ms}^{-2}$.

2 A resultant force of 100 N is applied to a body. Calculate the mass of the body when its acceleration is

(i) $0.5\,\text{ms}^{-2}$

(ii) $2\,\text{ms}^{-2}$

(iii) $0.01\,\text{ms}^{-2}$

(iv) $10g$.

3 What is the reaction between a book of mass 0.8 kg and your hand when it is

(i) accelerating downwards at $0.3\,\text{ms}^{-2}$

(ii) moving upwards at constant speed.

EXAMPLE 4.1

A lift and its passengers have a total mass of 400 kg. Find the tension in the cable supporting the lift when

(i) the lift is at rest
(ii) the lift is moving at constant speed
(iii) the lift is accelerating upwards at 0.8 ms^{-2}
(iv) the lift is accelerating downwards at 0.6 ms^{-2}.

SOLUTION

Before starting the calculations you must define a direction as positive. In this example the upward direction is chosen to be positive.

(i) At rest

As the lift is at rest the forces must be in equilibrium. The equation of motion is

$$T - mg = 0$$
$$T - 400 \times 9.8 = 0$$
$$T = 3920$$

The tension in the cable is 3920 N.

(ii) Moving at constant speed

Again, the forces on the lift must be in equilibrium because it is moving at a constant speed, so the tension is 3920 N.

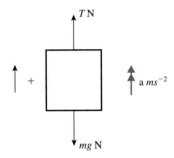

Figure 4.2

(iii) Accelerating upwards

The resultant upward force on the lift is $T - mg$ so the equation of motion is

$$T - mg = ma$$

which in this case gives

$$T - 400 \times 9.8 = 400 \times 0.8$$
$$T - 3920 = 320$$
$$T = 4240$$

The tension in the cable is 4240 N.

(iv) Accelerating downwards

The equation of motion is

$$T - mg = ma$$

In this case, a is negative so

$$T - 400 \times 9.8 = 400 \times (-0.6)$$
$$T - 3920 = -240$$
$$T = 3680$$

A downward acceleration of 0.6 ms^{-2} is an upward acceleration of -0.6 ms^{-2}

How is it possible for the tension to be 3680 N upwards but the lift to accelerate downwards?

EXAMPLE 4.2

This example shows how the *suvat* equations for motion with constant acceleration, which you met in Chapter 2, can be used with Newton's second law.

A supertanker of mass 500 000 tonnes is travelling at a speed of 10 ms^{-1} when its engines fail. It then takes half an hour for the supertanker to stop.

(i) Find the force of resistance, assuming it to be constant, acting on the supertanker.

When the engines have been repaired it takes the supertanker 10 minutes to return to its full speed of 10 ms^{-1}.

(ii) Find the driving force produced by the engines, assuming this also to be constant.

SOLUTION

Use the direction of motion as positive.

(i) First find the acceleration of the supertanker, which is constant for constant forces. Figure 4.3 shows the velocities and acceleration.

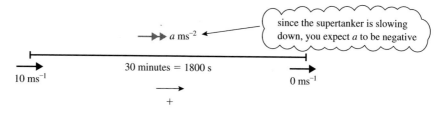

a ms^{-2} since the supertanker is slowing down, you expect a to be negative

30 minutes = 1800 s

10 ms^{-1} 0 ms^{-1}

+

Figure 4.3

You know $u = 10$, $v = 0$, $t = 1800$ and you want a, so use $v = u + at$

$$0 = 10 + 1800a$$

$$a = -\frac{1}{180}$$

The acceleration is negative because the supertanker is slowing down

Now we can use Newton's second law (Newton II) to write down the equation of motion. Figure 4.4 shows the horizontal forces and the acceleration.

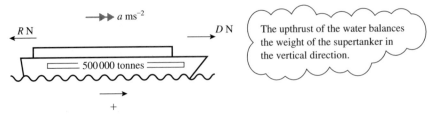

Figure 4.4

The resultant forwards force is $D - R$ newtons. When there is no driving force $D = 0$ so Newton II gives

> the mass must be in kg

$$0 - R = 500\,000\,000 \times a$$

so when $a = -\frac{1}{180}$, $-R = 500\,000\,000 \times \left(-\frac{1}{180}\right)$

The resistance to motion is 2.78×10^6 N or 2780 kN (correct to 3 sf).

 You have to be very careful with signs here: the resultant force and acceleration are both positive towards the right.

(ii) Now $u = 0$, $v = 10$ and $t = 600$, and we want a, so use $v = u + at$ again

$$10 = 0 + a \times 600$$

$$a = \frac{1}{60}$$

Using Newton's second law again

$$D - R = 500\,000\,000 \times a$$

$$D - 2.78 \times 10^6 = 500\,000\,000 \times \frac{1}{60}$$

$$D = 2.78 \times 10^6 + 8.33 \times 10^6$$

The driving force is 11.11×10^6 N or $11\,100$ kN (correct to 3 sf).

Tackling mechanics problems

When you tackle mechanics problems such as these you will find them easier if you:

- always draw a clear diagram
- clearly indicate the positive direction
- label each object (A, B, etc. or whatever is appropriate)
- show all the forces acting on each object
- make it clear which object you are referring to when writing an equation of motion.

1 A man pushes a car of mass 400 kg on level ground with a force of 200 N. The car is initially at rest and the man maintains this force until the car reaches a speed of 5 ms^{-1}. Ignoring any resistances forces, find
   (i) the acceleration of the car
   (ii) the distance the car travels while the man is pushing.

2 The engine of a car of mass 1.2 tonnes can produce a driving force of 2000 N. Ignoring any resistance forces, find
   (i) the car's resulting acceleration
   (ii) the time taken for the car to go from rest to 27 ms^{-1} (about 60 mph).

3 A top sprinter of mass 65 kg starting from rest reaches a speed of 10 ms^{-1} in 2 s.
   (i) Calculate the force required to produce this acceleration, assuming it is uniform.
   (ii) Compare this to the force exerted by a weight lifter holding a mass of 180 kg above the ground.

4 An ice skater of mass 65 kg is initially moving with speed 2 ms^{-1} and glides to a halt over a distance of 10 m. Assuming that the force of resistance is constant, find
   (i) the size of the resistance force
   (ii) the distance he would travel gliding to rest from an initial speed of 6 ms^{-1}
   (iii) the force he would need to apply to maintain a steady speed of 10 ms^{-1}.

5 A helicopter of mass 1000 kg is taking off vertically.
   (i) Draw a labelled diagram showing the forces on the helicopter as it lifts off and the direction of its acceleration.
   (ii) Its initial upward acceleration is 1.5 ms^{-2}. Calculate the upward force its rotors exert. Ignore the effects of air resistance.

6 Pat and Nicholas are controlling the movement of a canal barge by means of long ropes attached to each end. The tension in the ropes may be assumed to be horizontal and parallel to the line and direction of motion of the barge, as shown in the diagrams.

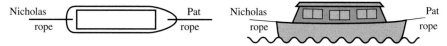

Plan   Elevation

The mass of the barge is 12 tonnes and the total resistance to forward motion may be taken to be 250 N at all times. Initially Pat pulls the barge forwards from rest with a force of 400 N and Nicholas leaves his rope slack.
   (i) Write down the equation of motion for the barge and hence calculate its acceleration.

Pat continues to pull with the same force until the barge has moved 10 m.
   (ii) What is the speed of the barge at this time and for what length of time did Pat pull?

Pat now lets her rope go slack and Nicholas brings the barge to rest by pulling with a constant force of 150 N.

(iii) Calculate

(a) how long it takes the barge to come to rest

(b) the total distance travelled by the barge from when it first moved

(c) the total time taken for the motion. [MEI]

7 A spaceship of mass 5000 kg is stationary in deep space. It fires its engines, producing a forward thrust of 2000 N for 2.5 minutes, and then turns them off.

(i) What is the speed of the spaceship at the end of the 2.5 minute period?

(ii) Describe the subsequent motion of the spaceship.

The spaceship then enters a cloud of interstellar dust which brings it to a halt after a further distance of 7200 km.

(iii) What is the force of resistance (assumed constant) on the spaceship from the interstellar dust cloud?

The spaceship is travelling in convoy with another spaceship which is the same in all respects except that it is carrying an extra 500 kg of equipment. The second spaceship carries out exactly the same procedure as the first one.

(iv) Which spaceship travels further into the dust cloud?

8 A crane is used to lift a hopper full of cement to a height of 20 m on a building site. The hopper has mass 200 kg and the cement 500 kg. Initially the hopper accelerates upwards at 0.05 ms^{-2}, then it travels at constant speed for some time before decelerating at 0.1 ms^{-2} until it is at rest. The hopper is then emptied.

(i) Find the tension in the crane's cable during each of the three phases of the motion and after emptying.

The cable's maximum safe load is 10 000 N.

(ii) What is the greatest mass of cement that can safely be transported in the same manner?

The cable is in fact faulty and on a later occasion breaks without the hopper leaving the ground. On that occasion the hopper is loaded with 720 kg of cement.

(iii) What can you say about the strength of the cable?

9 The police estimate that for good road conditions the frictional force, **F**, on a skidding vehicle of mass m is given by $\mathbf{F} = 0.8\,mg$. A car of mass 450 kg skids to a halt narrowly missing a child. The police measure the skid marks and find they are 12.0 m long.

(i) Calculate the deceleration of the car when it was skidding to a halt.

The child's mother says the car was travelling well over the speed limit but the driver of the car says she was travelling at 30 mph and the child ran out in front of her. (There are about 1600 m in a mile.)

(ii) Calculate the speed of the car when it started to skid.

Who was telling the truth?

Newton's second law applied to connected objects

This section is about using Newton's second law for more than one object. It is important to be very clear which forces act on which object in these cases.

A stationary helicopter is raising two people of masses 90 kg and 70 kg as shown in the diagram.

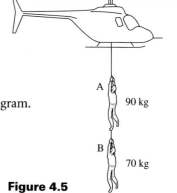

A 90 kg

B 70 kg

Figure 4.5

? Imagine that you are each person in turn. Your eyes are shut so you cannot see the helicopter or the other person. What forces act on you?

Remember that all the forces acting, apart from your weight, are due to contact between you and something else.

Which forces acting on A and B are equal in magnitude? What can you say about their accelerations?

EXAMPLE 4.3

(i) Draw a diagram to show the forces acting on the two people being raised by the helicopter in figure 4.5 and their acceleration.
(ii) Write down the equation of motion for each person.
(iii) When the force applied to the first person, A, by the helicopter is $180g$ N, calculate
  (a) the acceleration of the two people being raised
  (b) the tension in the ropes.
  Use $10\ \text{ms}^{-2}$ for g.

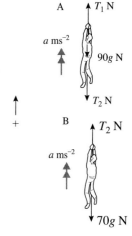

A T_1 N

$a\ \text{ms}^{-2}$

$90g$ N

T_2 N

+

B T_2 N

$a\ \text{ms}^{-2}$

$70g$ N

Figure 4.6

SOLUTION

(i) Figure 4.6 shows the acceleration and forces acting on the two people.
(ii) When the helicopter applies a force T_1 N to A, the resultant upward forces are
  A $(T_1 - 90g - T_2)\ \text{N}$ B $T_2 - 70g$
  Their equations of motion are
  A ($\uparrow$) $T_1 - 90g - T_2 = 90a$ ① B ($\uparrow$) $T_2 - 70g = 70a$ ②

(iii) You can eliminate T_2 from equations ① and ② by adding:

$$T_1 - 90g - T_2 + T_2 - 70g = 160a$$
$$T_1 - 160g = 160a \qquad \text{③}$$

When the force applied by the helicopter is $T_1 = 180g$

$$20g = 160a$$
$$a = 1.25$$

Substituting for a in equation ② gives $\qquad T_2 = 70 \times 1.25 + 70g$
$$= 787.5$$

The acceleration is $1.25 \, \text{ms}^{-2}$ and the tensions in the ropes are 1800 N and 787.5 N.

? The force pulling downwards on A is 787.5 N. This is *not* equal to B's weight (700 N). Why are they different?

Treating the system as a whole

When two objects are moving in the same direction with the same velocity at all times they can be treated as one. In example 4.3 the two people can be treated as one object and then the equal and opposite forces T_2 cancel out. They are *internal forces* similar to the forces between your head and your body.

The resultant upward force on both people is $T_1 - 90g - 70g$ and the total mass is 160 kg so the equation of motion is:

$$T_1 - 90g - 70g = 160a \quad \longleftarrow \quad \text{as equation ③ above}$$

So you can find a directly

$$\text{when } T_1 = 180g$$
$$20g = 160a$$
$$a = 1.25 \text{ when } g = 10$$

Treating the system as a whole finds a, but not the internal force T_2.

You need to consider the motion of B separately to obtain equation ②.

$$T_2 - 70g = 70a \qquad \text{②}$$
$$T_2 = 787.5 \quad \longleftarrow \quad \text{as before}$$

Using this method, equation ① can be used to check your answers. Alternatively, you could use equation ① to find T_2 and equation ② to check your answers.

When several objects are joined there are always more equations possible than are necessary to solve a problem and they are not all independent. In the above example, only two of the equations were necessary to solve the problem. The trick is to choose the most relevant ones.

A note on mathematical modelling

Several modelling assumptions have been made in the solution to example 4.3. It is assumed that:

- the only forces acting on the people are their weights and the tensions in the ropes. Forces due to the wind or air turbulence are ignored.
- the motion is vertical and nobody swings from side to side.
- the ropes do not stretch (i.e. they are inextensible) so the accelerations of the two people are equal.
- the people are rigid bodies which do not change shape and can be treated as particles.

All these modelling assumptions make the problem simpler. In reality, if you were trying to solve such a problem you might work through it first using these assumptions. You would then go back and decide which ones needed to be modified to produce a more realistic solution.

In the next example one person is moving vertically and the other horizontally. You might find it easier to decide on which forces are acting if you imagine you are Alvin or Bernard and you can't see the other person.

EXAMPLE 4.4

Alvin is using a snowmobile to pull Bernard out of a crevasse. His rope passes over a smooth block of ice at the top of the crevasse as shown in figure 4.7 and Bernard hangs freely away from the side. Alvin and his snowmobile together have a mass of 300 kg and Bernard's mass is 75 kg. Ignore any resistance to motion.

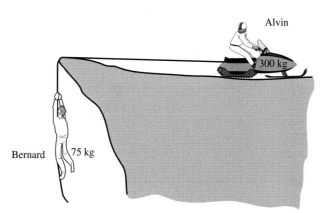

Figure 4.7

(i) Draw diagrams showing the forces on the snowmobile (including Alvin) and on Bernard.
(ii) Calculate the driving force required for the snowmobile to give Bernard an upward acceleration of $0.5\,\text{ms}^{-2}$ and the tension in the rope for this acceleration.
(iii) How long will it take for Bernard's speed to reach $5\,\text{ms}^{-1}$ starting from rest and how far will he have been raised in this time?

SOLUTION

(i) The diagram shows the essential features of the problem.

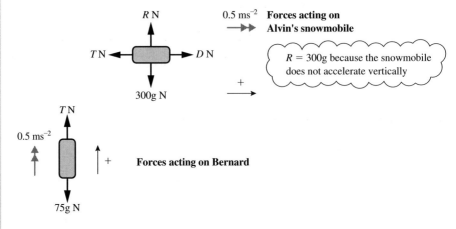

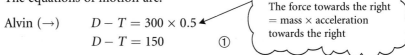

Forces acting on Bernard

Figure 4.8

(ii) Alvin and Bernard have the same acceleration providing the rope does not stretch. The tension in the rope is T newtons and Alvin's driving force is D newtons.

The equations of motion are:

Alvin ($\rightarrow$) $D - T = 300 \times 0.5$

$D - T = 150$ ①

The force towards the right = mass × acceleration towards the right

Bernard ($\uparrow$) $T - 75g = 75 \times 0.5$

$T - 75g = 37.5$ ②

$T = 37.5 + 75g$

$T = 772.5$

The upwards force = mass × upwards acceleration

Substituting in equation ①

$D - 772.5 = 150$

$D = 922.5$

The driving force required is 922.5 N and the tension in the rope is 772.5 N.

(iii) When $u = 0$, $v = 5$, $a = 0.5$ and t is required

$v = u + at$

$5 = 0 + 0.5 \times t$

$t = 10$

The time taken is 10 seconds.

Then using $s = ut + \frac{1}{2}at^2$ to find s gives

$s = 0 + \frac{1}{2}at^2$

$v^2 = u^2 + 2as$ would also give s

$s = \frac{1}{2} \times 0.5 \times 100$

$s = 25$

? Alvin thinks the rope will not stand a tension of more than 1.2 kN. What is the maximum safe acceleration in this case? Under the circumstances, is Alvin likely to use this acceleration?

Make a list of the modelling assumptions made in this example and suggest what effect a change in each of these assumptions might have on the solution.

EXAMPLE 4.5

A woman of mass 60 kg is standing in a lift.

(i) Draw a diagram showing the forces acting on the woman.

Find the normal reaction of the floor of the lift on the woman in the following cases.

(ii) The lift is moving upwards at a constant speed of 3 ms⁻¹.
(iii) The lift is moving upwards with an acceleration of 2 ms⁻² upwards.
(iv) The lift is moving downwards with an acceleration of 2 ms⁻² downwards.
(v) The lift is moving downwards and slowing down with a deceleration of 2 ms⁻².

In order to calculate the maximum number of occupants that can safely be carried in the lift, the following assumptions are made:

The lift has mass 300 kg, all resistances to motion may be neglected, the mass of each occupant is 75 kg and the tension in the supporting cable should not exceed 12 000 N.

(vi) What is the greatest number of occupants that can be carried safely if the magnitude of the acceleration does not exceed 3 ms⁻²?

[MEI]

SOLUTION

(i) The diagram shows the forces acting on the woman and her acceleration.

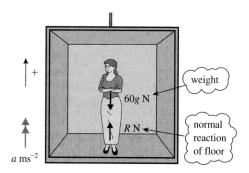

Figure 4.9

In general, when positive is upwards, her equation of motion is

(↑) $R - 60g = 60a$ ◄

> This equation contains all the mathematics in the situation. It can be used to solve parts (ii) to (iv)

(ii) When the speed is constant $a = 0$ so $R = 60g = 588.$ ← $g = 9.8$

The normal reaction is 588 N.

(iii) When $a = 2$

$$R - 60g = 60 \times 2$$
$$R = 120 + 588$$
$$= 708$$

The normal reaction is 708 N.

(iv) When the acceleration is downwards, $a = -2$ so

$$R - 60g = 60 \times (-2)$$
$$R = 468$$

The normal reaction is 468 N.

(v) When the lift is moving downwards and slowing down, the acceleration is negative downwards, so it is positive upwards, and $a = +2$. Then $R = 708$ as in part (iii).

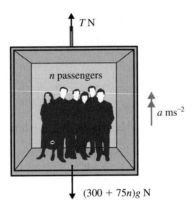

Figure 4.10

$(300 + 75n)g$ N

(vi) When there are n passengers in the lift, the combined mass of these and the lift is $(300 + 75n)$ kg and their weight is $(300 + 75n)g$ N.

The equation of motion for the lift and passengers together is

$$T - (300 + 75n)g = (300 + 75n)\, a$$

So when $a = 3$ and $g = 9.8$, $\quad T = (300 + 75n) \times 3 + (300 + 75n) \times 9.8$
$$= 12.8\,(300 + 75n)$$

For a maximum tension of 12 000 N

$$12\,000 = 12.8\,(300 + 75n)$$
$$937.5 = 300 + 75n$$
$$637.5 = 75n$$
$$n = 8.5$$

The lift cannot carry more than 8 passengers.

Remember: Always make it clear which object each equation of motion refers to.

1 Masses A of 100 g and B of 200 g are attached to the ends of a light, inextensible string which hangs over a smooth pulley as shown in the diagram.

Initially B is held at rest 2 m above the ground and A rests on the ground with the string taut. Then B is let go.

(i) Draw a diagram for each mass showing the forces acting on it and the direction of its acceleration at a later time when A and B are moving with an acceleration of $a\,\text{ms}^{-2}$ and before B hits the ground.

(ii) Write down the equation of motion of each mass in the direction it moves using Newton's second law.

(iii) Use your equations to find a and the tension in the string.

(iv) Find the time taken for B to hit the ground.

2 In this question you should take g to be $10\,\text{ms}^{-2}$. The diagram shows a block of mass 5 kg lying on a smooth table. It is attached to blocks of mass 2 kg and 3 kg by strings which pass over smooth pulleys. The tensions in the strings are T_1 and T_2, as shown, and the blocks have acceleration $a\,\text{ms}^{-2}$.

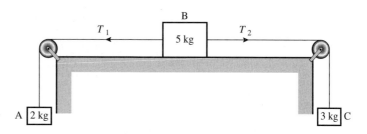

(i) Draw a diagram for each block showing all the forces acting on it and its acceleration.

(ii) Write down the equation of motion for each of the blocks.

(iii) Use your equations to find the values of a, T_1 and T_2.

In practice, the table is not truly smooth and a is found to be $0.5\,\text{ms}^{-2}$.

(iv) Repeat parts (i) and (ii) including a frictional force on B and use your new equations to find the frictional force that would produce this result.

3 A car of mass 800 kg is pulling a caravan of mass 1000 kg along a straight, horizontal road. The caravan is connected to the car by means of a light, rigid tow bar. The car is exerting a driving force of 1270 N. The resistances to the forward motion of the car and caravan are 400 N and 600 N respectively; you may assume that these resistances remain constant.

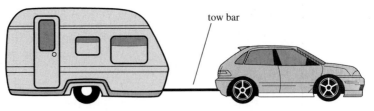

(i) Show that the acceleration of the car and caravan is $0.15\,\text{ms}^{-2}$.

(ii) Draw a diagram showing all the forces acting on the caravan along the line of its motion. Calculate the tension in the tow bar.

The driving force is removed but the car's brakes are not applied.

(iii) Determine whether the tow bar is now in tension or compression.

The car's brakes are then applied gradually. The brakes of the caravan come on automatically when the tow bar is subjected to a compression force of at least 50 N.

(iv) Show that the acceleration of the caravan just before its brakes come on automatically is $-0.65\,\text{ms}^{-2}$ in the direction of its motion. Hence, calculate the braking force on the car necessary to make the caravan brakes come on.

[MEI]

4 The diagram shows a goods train consisting of an engine of mass 40 tonnes and two trucks of 20 tonnes each. The engine is producing a driving force of 5×10^4 N, causing the train to accelerate. The ground is level and resistance forces may be neglected.

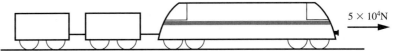

(i) By considering the motion of the whole train, find its acceleration.

(ii) Draw a diagram to show the forces acting on the engine and use this to help you to find the tension in the first coupling.

(iii) Find the tension in the second coupling.

The brakes on the first truck are faulty and suddenly engage, causing a resistance of 10^4 N.

(iv) What effect does this have on the tension in the coupling to the last truck?

[MEI, adapted]

5

A short train consists of two locomotives, each of mass 20 tonnes, with a truck of mass 10 tonnes coupled between them, as shown in the diagram. The resistances to forward motion are 0.5 kN on the truck and 1 kN on each of the locomotives. The train is travelling along a straight, horizontal section of track.

Initially there is a driving force of 15 kN from the front locomotive only.

(i) Calculate the acceleration of the train.

(ii) Draw a diagram indicating the horizontal forces acting on each part of the train, including the forces in each of the couplings. Calculate the forces acting on the truck due to each coupling.

On another occasion each of the locomotives produces a driving force of 7.5 kN in the same direction and the resistances remain as before.

(iii) Find the acceleration of the train and the forces now acting on the truck due to each of the couplings. Compare your answer to this part with your answer to part (ii) and comment briefly.

[MEI]

6 The diagram shows a lift containing a single passenger.

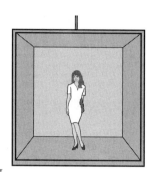

(i) Make clear diagrams to show the forces acting on the passenger and the forces acting on the lift using the following letters:

the tension in the cable, T N
the reaction of the lift on the
   passenger, R_P N
the reaction of the passenger on the lift, R_L N
the weight of the passenger, mg N
the weight of the lift, Mg N.

The masses of the lift and the passenger are 450 kg and 50 kg respectively.

(ii) Calculate T, R_P and R_L when the lift is stationary.

The lift then accelerates upwards at $0.8 \, \text{ms}^{-2}$.

(iii) Find the new values of T, R_P and R_L.

7 A man of mass 70 kg is standing in a lift which has an upward acceleration $a \, \text{ms}^{-2}$.

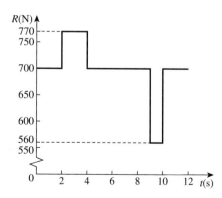

(i) Draw a diagram showing the man's weight, the force, R N, that the lift floor exerts on him and the direction of his acceleration.

(ii) Taking g to be $10 \, \text{ms}^{-2}$ find the value of a when $R = 770$ N.

The graph shows the value of R from the time ($t = 0$) when the man steps into the lift to the time ($t = 12$) when he steps out.

(iii) Explain what is happening in each section of the journey.

(iv) Draw the corresponding speed–time graph.

(v) To what height does the man ascend?

8 A lift in a mine shaft takes exactly one minute to descend 500 m. It starts from rest, accelerates uniformly for 12.5 seconds to a constant speed which it maintains for some time and then decelerates uniformly to stop at the bottom of the shaft.

The mass of the lift is 5 tonnes and on the day in question it is carrying 12 miners whose average mass is 80 kg.

(i) Sketch the speed–time graph of the lift.

During the first stage of the motion the tension in the cable is 53 640 N.

(ii) Find the acceleration of the lift during this stage.

(iii) Find the length of time for which the lift is travelling at constant speed and find the final deceleration.

(iv) What is the maximum value of the tension in the cable?

(v) Just before the lift stops one miner experiences an upthrust of 1002 N from the floor of the lift. What is the mass of the miner?

Reviewing a mathematical model: air resistance

? Why does a leaf or a feather or a piece of paper fall more slowly than other objects?

The model you have used so far for falling objects has assumed no air resistance and this is clearly unrealistic in many circumstances. There are several possible models for air resistance but it is usually better when modelling to try simple models first. Having rejected the first model you could try a second one as follows.

Model 2: Air resistance is constant and the same for all objects.

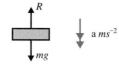

Figure 4.11

Assume an object of mass m falls vertically through the air.

The equation of motion is $mg - R = ma$

$$a = g - \frac{R}{m}$$

The model predicts that a heavy object will have a greater acceleration than a lighter one because $\frac{R}{m}$ is smaller for larger m.

This seems to agree with our experience of dropping a piece of paper and a book, for example. The heavier book has a greater acceleration.

? However, think again about air resistance. Is there a property of the object other than its mass which might affect its motion as it falls? How do people and animals maximise or minimise the force of the air?

Try dropping two identical sheets of paper from a horizontal position, but fold one of them. The folded one lands first even though they have the same mass.

This contradicts the prediction of model 2. A large surface at right angles to the motion seems to increase the resistance.

Model 3: Air resistance is proportional to the area perpendicular to the motion.

Assume the air resistance is kA where k is constant and A is the area of the surface perpendicular to the motion.

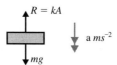

Figure 4.12

The equation of motion is now $mg - kA = ma$

$$a = g - \frac{kA}{m}$$

According to this model, the acceleration depends on the ratio of the area to the mass.

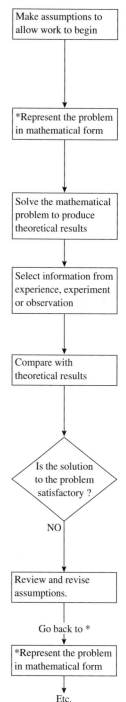

Steps in the modelling procedure

Make assumptions to allow work to begin

*Represent the problem in mathematical form

Solve the mathematical problem to produce theoretical results

Select information from experience, experiment or observation

Compare with theoretical results

Is the solution to the problem satisfactory ?

NO

Review and revise assumptions.

Go back to *

*Represent the problem in mathematical form

Etc.

TESTING THE NEW MODEL

For this experiment you will need some rigid corrugated card such as that used for packing or in grocery boxes (cereal box card is too thin), scissors and tape.

Cut out ten equal squares of side 8 cm. Stick two together by binding the edges with tape to make them smooth. Then stick three and four together in the same way so that you have four blocks A to D of different thickness as shown in the diagram.

Cut out ten larger squares with 12 cm sides. Stick them together in the same way to make four blocks E to H.

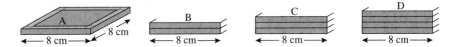

Figure 4.13

Observe what happens when you hold one or two blocks horizontally at a height of about 2 m and let them fall. You do not need to measure anything in this experiment, unless you want to record the area and mass of each block, but write down your observations in an orderly fashion.

1 Drop each one separately. Could its acceleration be constant?
2 Compare A with B and B with D. Make sure you drop each pair from the same height and at the same instant of time. Do they take the same time to fall? Predict what will happen with other combinations and test your predictions.
3 Experiment in a similar way with E to H.
4 Now compare A with E; B with F; C with G and D with H.
   Compare also the two blocks whose dimensions are all in the same ratio, i.e. B and G.

? Do your results suggest that model 3 might be better than model 2?

If you want to be more certain, the next step would be to make accurate measurements. Nevertheless, this model explains why small animals can be relatively unscathed after falling through heights which would cause serious injury to human beings.

? All the above models ignore one important aspect of air resistance. What is that?

1 **The equation of motion**

Newton's second law gives *the equation of motion* for an object.

$$\text{The resultant force} = \text{mass} \times \text{acceleration} \quad \text{or} \quad \mathbf{F} = m\mathbf{a}$$

The acceleration is always in the same direction as the resultant force.

2 **Connected objects**

- Reaction forces between two objects (such as tension forces in joining rods or strings) are equal and opposite.

- When connected objects are moving along a line, the equations of motion can be obtained for each one separately or for a system containing more than one object. The number of independent equations is equal to the number of separate objects.

3 **Reviewing a model**

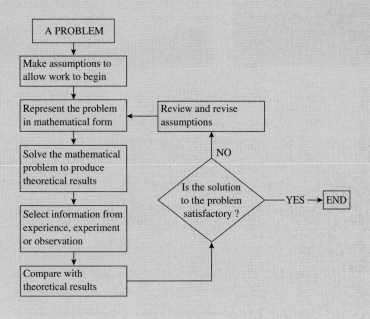

5 Vectors

But the principal failing occurred in the sailing
And the bellman, perplexed and distressed,
Said he had hoped, at least when the wind blew due East
That the ship would not travel due West.

Lewis Carroll

Adding vectors

? If you walk 12 m east and then 5 m north, how far and in what direction will you be from your starting point?

A bird is caught in a wind blowing east at $12\,\text{ms}^{-1}$ and flies so that its speed would be $5\,\text{ms}^{-1}$ north in still air. What is its actual velocity?

A sledge is being pulled with forces of 12 N east and 5 N north. What single force would have the same effect?

All these questions involve vectors – displacement, velocity and force. When you are concerned only with the *magnitude and direction* of these vectors, the three problems can be reduced to one. They can all be solved using the same vector techniques.

Displacement vectors

The instruction 'walk 12 m east and then 5 m north' can be modelled mathematically using a scale diagram, as in figure 5.1. The arrowed lines AB and BC are examples of vectors.

We write the vectors as $\overrightarrow{AB}$ and $\overrightarrow{BC}$. The arrow above the letters is very important as it indicates the direction of the vector. $\overrightarrow{AB}$ means from A to B. $\overrightarrow{AB}$ and $\overrightarrow{BC}$ are examples of *displacement vectors*. Their lengths represent the magnitude of the displacements.

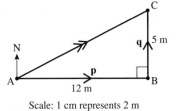

Scale: 1 cm represents 2 m

Figure 5.1

It is often more convenient to use a single letter to denote a vector. For example in textbooks and exam papers you might see the displacement vectors $\overrightarrow{AB}$ and $\overrightarrow{BC}$ written as **p** and **q** (i.e. in bold print). When writing these vectors yourself, you should underline your letters, e.g. p̲ and q̲.

The magnitudes of **p** and **q** are then shown as $|\mathbf{p}|$ and $|\mathbf{q}|$ or p and q (in ordinary print).
These are scalar quantities.

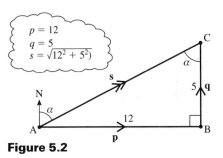

Figure 5.2

The combined effect of the two displacements $\overrightarrow{AB}$ (=**p**) and $\overrightarrow{BC}$ (=**q**) is $\overrightarrow{AC}$ and this is called the *resultant vector*. It is marked with two arrows to distinguish it from **p** and **q**. The process of combining vectors in this way is called *vector addition*. We write $\overrightarrow{AB} + \overrightarrow{BC} = \overrightarrow{AC}$ or $\mathbf{p} + \mathbf{q} = \mathbf{s}$.

You can calculate the resultant using Pythagoras' theorem and trigonometry.

In triangle ABC $\quad AC = \sqrt{12^2 + 5^2} = 13$

and $\qquad\qquad \tan \alpha = \dfrac{12}{5}$

$\qquad\qquad\quad \alpha = 67°$ (to the nearest degree)

The distance from the starting point is 13 m and the direction is 067°.

Velocity and force

The other two problems that begin this chapter are illustrated in these diagrams.

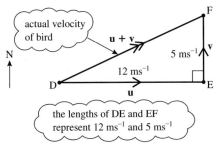

Figure 5.3

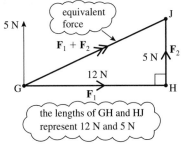

Figure 5.4

When $\overrightarrow{DE}$ represents the velocity (**u**) of the wind and $\overrightarrow{EF}$ represents the velocity (**v**) of the bird in still air, the vector $\overrightarrow{DF}$ represents the resultant velocity, **u** + **v**.

 Why does the bird move in the direction DF? Think what happens in very small intervals of time.

In figure 5.4, the vector $\overrightarrow{GJ}$ represents the equivalent (resultant) force. You know that it acts at the same point as the children's forces, but its magnitude and direction can be found using the triangle GHJ which is similar to the two triangles, ABC and DEF.

The same diagram does for all, you just have to supply the units. The bird travels at 13 ms^{-1} in the direction of 067° and one child would have the same effect as the others by pulling with a force of 13 N in the direction 067°. In most of this chapter vectors are treated in the abstract. You can then apply what you learn to different real situations.

Free vectors

Free vectors have magnitude and direction only. All vectors which are in the same direction and have the same magnitude are equal.

If the vector **p** represents a velocity of 3 km h^{-1} north-east, what do –**p** and 2**p** represent?

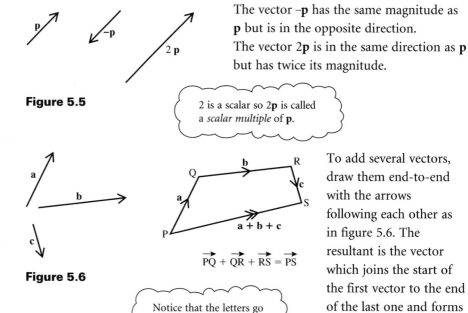

The vector –**p** has the same magnitude as **p** but is in the opposite direction.
The vector 2**p** is in the same direction as **p** but has twice its magnitude.

Figure 5.5

> 2 is a scalar so 2**p** is called a *scalar multiple* of **p**.

Figure 5.6

> Notice that the letters go together like dominoes.

$$\overrightarrow{PQ} + \overrightarrow{QR} + \overrightarrow{RS} = \overrightarrow{PS}$$

a + b + c

To add several vectors, draw them end-to-end with the arrows following each other as in figure 5.6. The resultant is the vector which joins the start of the first vector to the end of the last one and forms the last side of a polygon. Notice its direction.

EXAMPLE 5.1

The diagram shows several vectors.

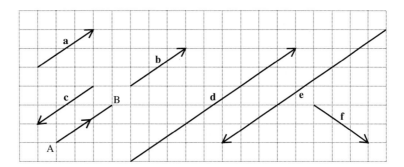

Figure 5.7

(i) Write each of the other vectors in terms of the vector **a**.
(ii) Draw scale diagrams to show
    (a) $\mathbf{a} + \mathbf{f}$ **(b)** $\mathbf{a} - \mathbf{f}$ **(c)** $2\mathbf{c} + \mathbf{f}$ **(d)** $\mathbf{a} + \mathbf{f} + \mathbf{c}$

SOLUTION

(i) $\mathbf{b} = \mathbf{a}$, $\overrightarrow{AB} = \mathbf{a}$, $\mathbf{c} = -\mathbf{a}$, $\mathbf{d} = 3\mathbf{a}$, $\mathbf{e} = -3\mathbf{a}$.
    The vector **f** cannot be written in terms of **a** but $|\mathbf{f}| = |\mathbf{a}|$.
(ii) Using vector addition the solutions are as shown below.

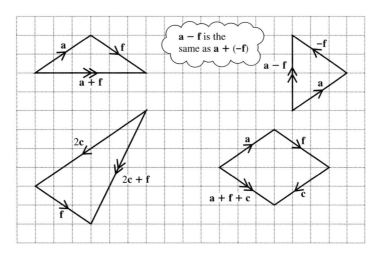

Figure 5.8

Adding parallel vectors

You can add parallel vectors by thinking of them as positive and negative, or by drawing diagrams as in Example 5.2.

EXAMPLE 5.2

The vector **p** is 10 units north-west and **q** is 6 units north-west.

(i) Describe the vector **p** − **q** and write the answer in terms of **p**.

(ii) Write **p** + **q** and **q** − **p** in terms of **p**.

(iii) The vector **s** is 5 units south-east. What is **p** + 2**s**?

SOLUTION

(i) The diagram shows the vectors **p**, **q** and **p** − **q**.

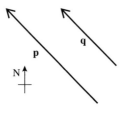
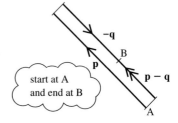

Figure 5.9

p − **q** is in the direction of **p** and of magnitude 10 − 6 = 4 units.
p − **q** = 0.4 **p**

(ii) **p** + **q** is (10 + 6) = 16 units NW so **p** + **q** = 1.6 **p**
q − **p** is (6 − 10) = −4 units NW or 4 units SE, so
q − **p** = −(**p** − **q**) = −0.4 **p**

(iii) **s** = −0.5 **p** so **p** + 2**s** = 0**p** = **0**

Note

We use **0** (in bold) and not 0 (in ordinary type) on the right-hand side of this expression to show that the quantity is still a vector.

EXERCISE 5A

Remember to underline all your vectors (these are printed in bold here).

1 The diagram shows several vectors.

Write each of the other vectors in terms of **a** or **b**. How are **a** and **b** related?

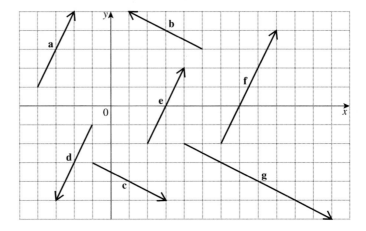

2 A child climbs up the ladder attached to a slide and then slides down. What three vectors model the displacement of the child during this activity?

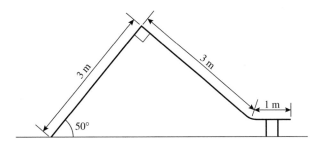

3 A child runs up and down a train. If the child runs at $2\,\text{ms}^{-1}$ and the train moves at $30\,\text{ms}^{-1}$, what are the resultant velocities of the child?

4 A girl rows at $5\,\text{ms}^{-1}$ in still water. What is her resultant velocity if she rows
   (i)   in the same direction as a current flowing at $3\,\text{ms}^{-1}$?
   (ii) in the opposite direction to the same current?
   (iii) in the opposite direction to a current flowing at $8\,\text{ms}^{-1}$?

5 A crane moves a crate from the ground 10 m vertically upward, then 6 m horizontally and 2 m vertically downward. Draw a scale diagram of the path of the crate. What single translation would move the crate to its final position from its initial position on the ground?

6 The three vectors **a**, **b** and **c** have magnitudes 1, 2 and 5 as shown in the diagram.
   (i) Draw scale diagrams to illustrate **a** + **b** and **b** + **a**. What do you conclude?
   (ii) Draw a scale diagram to illustrate **a** + **b** + **c**. Explain why you need do no more work to find **c** + **a** + **b**.

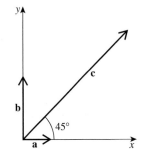

7 An irregular pentagon ABCDE is shown in the diagram. Write down relationships between the following vectors:
   (i) $\overrightarrow{AB}$ and $\overrightarrow{DC}$
   (ii) $\overrightarrow{AB}$ and $\overrightarrow{CD}$
   (iii) $\overrightarrow{EA}$ and $\overrightarrow{CB}$
   (iv) $\overrightarrow{AE}$ and $\overrightarrow{CB}$.

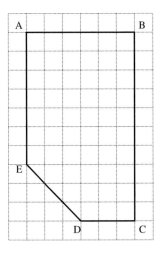

8 ABCD is a quadrilateral. Show that
$$\overrightarrow{AB} + \overrightarrow{BC} + \overrightarrow{CD} + \overrightarrow{DA} = \mathbf{0}.$$

9 The triangle ABC is such that $\overrightarrow{AB} = \mathbf{p}$ and $\overrightarrow{BC} = \mathbf{q}$.
Find expressions in terms of $\mathbf{p}$ and $\mathbf{q}$ for:

(i) $\overrightarrow{AC}$

(ii) $\overrightarrow{BP}$ where $\overrightarrow{BP} = 1/3\,\overrightarrow{BC}$

(iii) $\overrightarrow{AP}$

(iv) $\overrightarrow{AQ}$ where Q is the mid-point of AP.

10 In the parallelogram, $\overrightarrow{OA} = \mathbf{a}$,
$\overrightarrow{OC} = \mathbf{c}$, and M is the mid-point of AB.
Express the displacements below in
terms of $\mathbf{a}$ and $\mathbf{c}$:

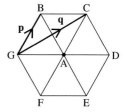

(i) $\overrightarrow{CB}$	**(ii)** $\overrightarrow{OB}$
(iii) $\overrightarrow{AC}$	**(iv)** $\overrightarrow{CA}$
(v) $\overrightarrow{BO}$	**(vi)** $\overrightarrow{AM}$
(vii) $\overrightarrow{OM}$	**(viii)** $\overrightarrow{MC}$.

11 In a regular hexagon, $\overrightarrow{GB} = \mathbf{p}$ and $\overrightarrow{GC} = \mathbf{q}$.
Express the following in terms of $\mathbf{p}$ and $\mathbf{q}$:

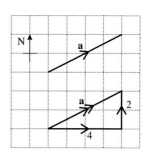

(i) $\overrightarrow{CA}$	**(ii)** $\overrightarrow{GA}$	**(iii)** $\overrightarrow{BC}$
(iv) $\overrightarrow{GF}$	**(v)** $\overrightarrow{GE}$	**(vi)** $\overrightarrow{CD}$.

Components of a vector

So far you have added two vectors to make one
resultant vector. Alternatively, it is often convenient
to write one vector in term of two others called
components.

The vector $\mathbf{a}$ in the diagram can be split into two
components in an infinite number of ways. All you
need to do is to make $\mathbf{a}$ one side of a triangle. It is
most sensible, however, to split vectors into
components in convenient directions and these
directions are usually perpendicular.

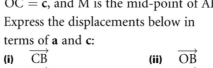

Figure 5.10

Using the given grid, $\mathbf{a}$ is 4 units east combined with 2 units north.

Unit vectors i and j

You can write $\mathbf{a}$ in figure 5.10 as $4\mathbf{i} + 2\mathbf{j}$ where $\mathbf{i}$ represents a vector of one unit
to the east and $\mathbf{j}$ a vector of one unit to the north. $\mathbf{i}$ and $\mathbf{j}$ are called *unit vectors*.

Alternatively $\mathbf{a}$ can be written as $\begin{pmatrix} 4 \\ 2 \end{pmatrix}$. This is called a *column vector*.

When using the standard Cartesian co-ordinate system, **i** is a vector of one unit along the *x* axis and **j** is a vector of one unit along the *y* axis. Any other vector drawn in the *xy* plane can then be written in terms of **i** and **j**.

You may define the unit vectors **i** and **j** to be in *any* two perpendicular directions if it is convenient to do so.

? What is **a** in terms of **i** and **j** if **i** is north-east and **j** is south-east?

> *Note*
>
> You have already worked with vectors in components. The total reaction between two surfaces is often split into two components. One (friction) is opposite to the direction of possible sliding and the other (normal reaction) is perpendicular to it.

EXAMPLE 5.3

The four vectors **a**, **b**, **c** and **d** are shown in the diagram.

(i) Write them in component form.

(ii) Draw a diagram to show 2**c** and –**d** and write them in component form.

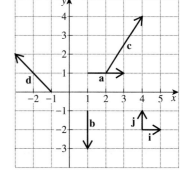

Figure 5.11

SOLUTION

(i) $\mathbf{a} = 2\mathbf{i} = \begin{pmatrix} 2 \\ 0 \end{pmatrix}$ $\mathbf{b} = -2\mathbf{j} = \begin{pmatrix} 0 \\ -2 \end{pmatrix}$

$\mathbf{c} = 2\mathbf{i} + 3\mathbf{j} = \begin{pmatrix} 2 \\ 3 \end{pmatrix}$ $\mathbf{d} = -2\mathbf{i} + 2\mathbf{j} = \begin{pmatrix} -2 \\ 2 \end{pmatrix}$

(ii) $2\mathbf{c} = 2(2\mathbf{i} + 3\mathbf{j})$ or $2\begin{pmatrix} 2 \\ 3 \end{pmatrix}$

$\qquad = 4\mathbf{i} + 6\mathbf{j}$ or $\begin{pmatrix} 4 \\ 6 \end{pmatrix}$

$-\mathbf{d} = -(-2\mathbf{i} + 2\mathbf{j})$ or $-\begin{pmatrix} -2 \\ 2 \end{pmatrix}$

$\qquad = 2\mathbf{i} - 2\mathbf{j}$ or $\begin{pmatrix} 2 \\ -2 \end{pmatrix}$

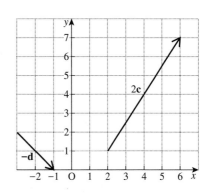

Figure 5.12

Equal vectors and parallel vectors

When two vectors, **p** and **q**, are *equal* then they must be equal in both magnitude and direction. If they are written in component form their components must be equal.

So if $\qquad$ $\mathbf{p} = a_1\,\mathbf{i} + b_1\,\mathbf{j}$

and $\qquad$ $\mathbf{q} = a_2\,\mathbf{i} + b_2\,\mathbf{j}$

then $\qquad$ $a_1 = a_2$ and $b_1 = b_2$.

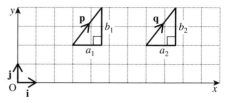

Figure 5.13

Thus in two dimensions, the statement $\mathbf{p} = \mathbf{q}$ is the equivalent of two equations (and in three dimensions, three equations).

If **p** and **q** are *parallel but not equal*, they make the same angle with the *x* axis.

then $\qquad$ $\dfrac{b_1}{a_1} = \dfrac{b_2}{a_2}$ or $\dfrac{a_1}{a_2} = \dfrac{b_1}{b_2}$

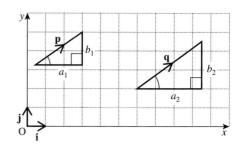

Figure 5.14

? If $\begin{pmatrix} 4 \\ 3 \end{pmatrix}$ is parallel to $\begin{pmatrix} -8 \\ y \end{pmatrix}$ what is *y*?

Position vectors

When an object is modelled as a particle or a point moving in space its *position* is its *displacement relative to a fixed origin*.

When O is the fixed origin, the vector $\overrightarrow{OA}$ (or **a**) is called the *position vector* of A.

If a point P has coordinates (3, 2) and **i** and **j** are in the direction of the *x* and *y* axes, the position of the point P is

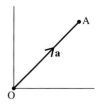

Figure 5.15

$$\overrightarrow{OP} = 3\mathbf{i} + 2\mathbf{j} \text{ or } \begin{pmatrix} 3 \\ 2 \end{pmatrix}.$$

Adding vectors in component form

In component form, addition and subtraction of vectors is simply carried out by adding or subtracting the components of the vectors.

EXAMPLE 5.4

Two vectors **a** and **b** are given by **a** = 2**i** + 3**j** and **b** = −**i** + 4**j**.

(ii) Find the vectors **a** + **b** and **a** − **b**.

(ii) Verify that your results are the same if you use a scale drawing.

SOLUTION

(i) Using **i** and **j**:

$$\mathbf{a} + \mathbf{b} = (2\mathbf{i} + 3\mathbf{j}) + (-\mathbf{i} + 4\mathbf{j})$$
$$= 2\mathbf{i} - \mathbf{i} + 3\mathbf{j} + 4\mathbf{j}$$
$$= \mathbf{i} + 7\mathbf{j}$$

$$\mathbf{a} - \mathbf{b} = (2\mathbf{i} + 3\mathbf{j}) - (-\mathbf{i} + 4\mathbf{j})$$
$$= 2\mathbf{i} + \mathbf{i} + 3\mathbf{j} - 4\mathbf{j}$$
$$= 3\mathbf{i} - \mathbf{j}$$

Using column vectors:

$$\mathbf{a} + \mathbf{b} = \begin{pmatrix} 2 \\ 3 \end{pmatrix} + \begin{pmatrix} -1 \\ 4 \end{pmatrix}$$
$$= \begin{pmatrix} 1 \\ 7 \end{pmatrix}$$

$$\mathbf{a} - \mathbf{b} = \begin{pmatrix} 2 \\ 3 \end{pmatrix} - \begin{pmatrix} -1 \\ 4 \end{pmatrix}$$
$$= \begin{pmatrix} 3 \\ -1 \end{pmatrix}$$

(ii)

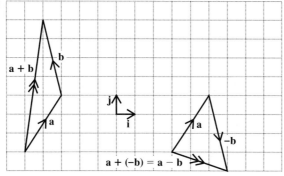

Figure 5.16

From the diagram you can see that $\mathbf{a} + \mathbf{b} = \mathbf{i} + 7\mathbf{j}$ or $\begin{pmatrix} 1 \\ 7 \end{pmatrix}$

and $\mathbf{a} - \mathbf{b} = 3\mathbf{i} - \mathbf{j}$ or $\begin{pmatrix} 3 \\ -1 \end{pmatrix}$.

These vectors are the same as those obtained in part (i).

? **a** and **b** are the position vectors of points A and B as shown in the diagram.

How can you write the displacement vector $\overrightarrow{AB}$ in terms of **a** and **b**?

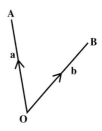

Figure 5.17

1 The diagram shows a grid of 1 m squares.
A person walks first east and then north.
How far should the person walk in each
of these directions to travel

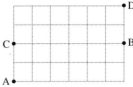
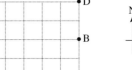

(i)   from A to B?

(ii) from B to C?

(iii) from A to D?

2 Write the vectors in the diagram in terms of unit vectors **i** and **j**.

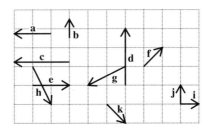

3 Given that $\mathbf{a} = \begin{pmatrix} 2 \\ -1 \end{pmatrix}$ and $\mathbf{b} = \begin{pmatrix} 1 \\ 4 \end{pmatrix}$ what are the coordinates of the point
with position vector $3\mathbf{a} - 2\mathbf{b}$?

4 Four vectors are given in component form by $\mathbf{a} = 3\mathbf{i} + 4\mathbf{j}$, $\mathbf{b} = 6\mathbf{i} - 7\mathbf{j}$,
$\mathbf{c} = -2\mathbf{i} + 5\mathbf{j}$ and $\mathbf{d} = -5\mathbf{i} - 3\mathbf{j}$.

Find the vectors:

(i) $\mathbf{a} + \mathbf{b}$

(ii) $\mathbf{b} + \mathbf{c}$

(iii) $\mathbf{c} + \mathbf{d}$

(iv) $\mathbf{a} + \mathbf{b} + \mathbf{d}$

(v) $\mathbf{a} - \mathbf{b}$

(vi) $\mathbf{d} - \mathbf{b} + \mathbf{a}$.

5 Given vectors $\mathbf{a} = \begin{pmatrix} 4 \\ 1 \end{pmatrix}$, $\mathbf{b} = \begin{pmatrix} -1 \\ 0 \end{pmatrix}$, $\mathbf{c} = \begin{pmatrix} -2 \\ -3 \end{pmatrix}$ and $\mathbf{d} = \begin{pmatrix} 2 \\ 6 \end{pmatrix}$, find

(i) $\mathbf{a} + 2\mathbf{b}$

(ii) $2\mathbf{c} - 3\mathbf{d}$

(iii) $\mathbf{a} + \mathbf{c} - 2\mathbf{b}$

(iv) $-2\mathbf{a} + 3\mathbf{b} + 4\mathbf{d}$.

6 A, B and C are the points (1, 2), (5, 1) and (7, 8).

(i)   Write down in terms of **i** and **j**, the position vectors of these three
points.

(ii) Find the component form of the displacements $\overrightarrow{AB}$, $\overrightarrow{BC}$ and $\overrightarrow{CA}$.

(iii) Draw a diagram to show the position vectors of A, B and C and your
answers to part (ii).

7 A, B and C are the points (0, –3), (2, 5) and (3, 9).

(i)   Write down in terms of **i** and **j** the position vectors of these three points.

(ii) Find the displacements $\overrightarrow{AB}$ and $\overrightarrow{BC}$.

(iii) Show that the three points all lie on a straight line.

8 A, B, C and D are the points (4, 2), (1, 3), (0, 10) and (3, d).

(i) Find the value of d so that DC is parallel to AB.

(ii) Find a relationship between $\overrightarrow{BC}$ and $\overrightarrow{AD}$. What is ABCD?

9   Three vectors **a**, **b** and **c** are given by $\mathbf{a} = \mathbf{i} + \mathbf{j}$, $\mathbf{b} = \mathbf{i} + 2\mathbf{j}$ and $\mathbf{c} = 3\mathbf{i} - 4\mathbf{j}$. R is the end-point of the displacement $2\mathbf{a} + 3\mathbf{b} + \mathbf{c}$ and (1, 2) is the starting point. What is the position vector of R?

10 Given the vectors $\mathbf{p} = 3\mathbf{i} - 5\mathbf{j}$ and $\mathbf{q} = -\mathbf{i} + 4\mathbf{j}$ find the vectors **x** and **y** where
   (i) $2\mathbf{x} - 3\mathbf{p} = \mathbf{q}$ (ii) $4\mathbf{p} - 3\mathbf{y} = 7\mathbf{q}$.

11 The vectors **x** and **y** are defined in terms of a and b as $\mathbf{x} = a\,\mathbf{i} + (a + b)\,\mathbf{j}$
   $\mathbf{y} = (6 - b)\,\mathbf{i} - (2a + 3)\,\mathbf{j}$.

   Given that $\mathbf{x} = \mathbf{y}$, find the values of a and b.

The magnitude and direction of vectors written in component form

At the beginning of this chapter the magnitude of a vector was found by using Pythagoras' theorem (see page 79). The direction was given using bearings, measured clockwise from the north.

When the vectors are in an xy plane, a mathematical convention is used for direction. Starting from the x axis, angles measured anticlockwise are positive and angles in a clockwise direction are negative as in figure 5.18.

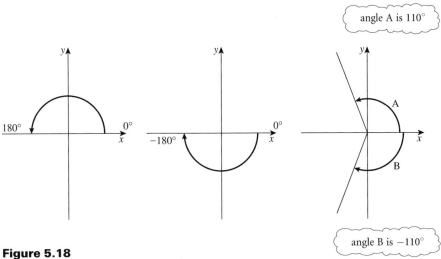

angle A is $110°$

angle B is $-110°$

Figure 5.18

Using the notation in figure 5.19, the magnitude and direction can be written in general form.

Magnitude of the vector $|a_1\,\mathbf{i} + a_2\,\mathbf{j}| = \sqrt{a_1{}^2 + a_2{}^2}$

Direction $\tan\theta = \dfrac{a_2}{a_1}$

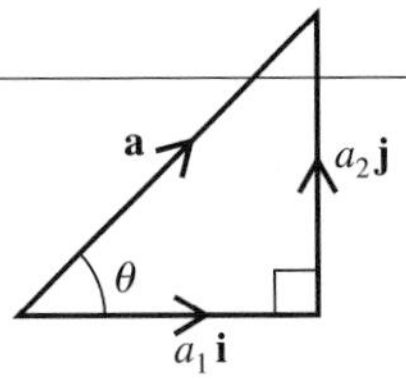

Figure 5.19

EXAMPLE 5.5

Find the magnitude and direction of the vectors $4\mathbf{i} + 3\mathbf{j}$, $4\mathbf{i} - 3\mathbf{j}$, $-4\mathbf{i} + 3\mathbf{j}$ and $-4\mathbf{i} - 3\mathbf{j}$ in the xy plane.

SOLUTION

First draw diagrams so that you can see which lengths and acute angles to find.

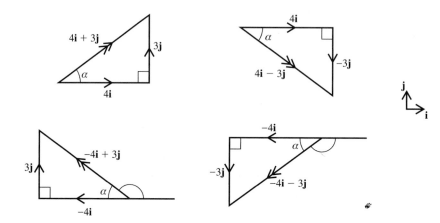

Figure 5.20

The vectors in each of the diagrams have the same magnitude and using Pythagoras' theorem, the resultants all have magnitude $\sqrt{4^2 + 3^2} = 5$.

The angles α are also the same size in each diagram and can be found using

$$\tan \alpha = \frac{3}{4}$$

$$\alpha = 37°.$$

The angles the vectors make with the $\mathbf{i}$ direction specify their directions:

$$
\begin{array}{ll}
4\mathbf{i} + 3\mathbf{j} & 37° \\
4\mathbf{i} - 3\mathbf{j} & -37° \\
-4\mathbf{i} + 3\mathbf{j} & 180° - 37° = 143° \\
-4\mathbf{i} - 3\mathbf{j} & -143°
\end{array}
$$

Unit vectors

Sometimes you need to write down a unit vector (i.e. a vector of magnitude 1) in the direction of a given vector. You have seen that the vector $a_1 \mathbf{i} + a_2 \mathbf{j}$ has magnitude $\sqrt{a_1^2 + a_2^2}$ and so it follows that the vector

$$\frac{a_1 \mathbf{i} + a_2 \mathbf{j}}{\sqrt{a_1^2 + a_2^2}} = \frac{a_1}{\sqrt{a_1^2 + a_2^2}} \mathbf{i} + \frac{a_2}{\sqrt{a_1^2 + a_2^2}} \mathbf{j} \text{ has magnitude 1.}$$

EXAMPLE 5.6

(i) Find a unit vector in the direction of $-4\mathbf{i} + 3\mathbf{j}$.

(ii) A force **F** has magnitude 8 N and is in the direction $-4\mathbf{i} + 3\mathbf{j}$. Write **F** in terms of **i** and **j**.

SOLUTION

(i) $|-4\mathbf{i} + 3\mathbf{j}| = \sqrt{4^2 + 3^2} = 5$

So a unit vector in this direction is

$$\frac{-4\mathbf{i} + 3\mathbf{j}}{5} = -0.8\mathbf{i} + 0.6\mathbf{j}.$$

Notice that $\sqrt{0.8^2 + 0.6^2} = 1$, so this is a unit vector.

Also $\dfrac{0.6}{-0.8} = -\dfrac{3}{4} = \tan 143°$ so it is

in the same direction as $-4\mathbf{i} + 3\mathbf{j}$.

(ii) **F** is 8 times the unit vector in the same direction, so

$$\mathbf{F} = 8(-0.8\mathbf{i} + 0.6\mathbf{j}) = -6.4\mathbf{i} + 4.8\mathbf{j}\,\text{N}.$$

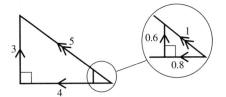

Figure 5.21

Vectors in three dimensions

In three dimensions there is a third, z, axis perpendicular to both the x and y axes through the origin 0 of the xy plane. When the x and y axes are drawn on a horizontal sheet, the z axis is drawn vertically upwards as shown in figure 5.22. Imagine turning the sheet so that the positive region faces you. The point Q $(2, 3, 6)$ would then be plotted as shown in figure 5.23.

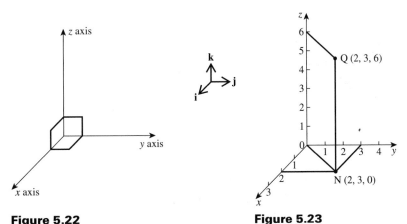

Figure 5.22   **Figure 5.23**

A third unit vector **k** is introduced to represent the positive z direction. For example, the position vector of the point Q $(2, 3, 6)$ is written as

$$\overrightarrow{OQ} = 2\mathbf{i} + 3\mathbf{j} + 6\mathbf{k} = \begin{pmatrix} 2 \\ 3 \\ 6 \end{pmatrix}.$$

In general the position vector of point P (x, y, z) is written as $\mathbf{r} = x\mathbf{i} + y\mathbf{j} + z\mathbf{k}$.

You can use Pythagoras' theorem to find the magnitude of a vector in space.

In figure 5.23 $OQ^2 = ON^2 + NQ^2$

$$= (2^2 + 3^2) + 6^2$$

$$OQ = \sqrt{2^2 + 3^2 + 6^2} = 7$$

A unit vector in the direction of OQ is $\frac{2}{7}\mathbf{i} + \frac{3}{7}\mathbf{j} + \frac{6}{7}\mathbf{k} = 0.29\mathbf{i} + 0.43\mathbf{j} + 0.86\mathbf{k}$.

In general, the vector $\mathbf{a} = a_1\mathbf{i} + a_2\mathbf{j} + a_3\mathbf{k}$ or $\begin{pmatrix} a_1 \\ a_2 \\ a_3 \end{pmatrix}$

has a magnitude of $\sqrt{a_1{}^2 + a_2{}^2 + a_3{}^2}$ and a unit vector in the direction of $\mathbf{a}$ is

$$\frac{a_1}{\sqrt{a_1{}^2 + a_2{}^2 + a_3{}^2}}\mathbf{i} + \frac{a_2}{\sqrt{a_1{}^2 + a_2{}^2 + a_3{}^2}}\mathbf{j} + \frac{a_3}{\sqrt{a_1{}^2 + a_2{}^2 + a_3{}^2}}\mathbf{k}.$$

EXERCISE 5C

Make use of sketches to help you in this exercise.

1 Find the magnitude and direction of each of these vectors.

(i) $\mathbf{a} = 2\mathbf{i} + 3\mathbf{j}$

(ii) $\mathbf{v} = 5\mathbf{i} - 12\mathbf{j}$

(iii) $\mathbf{F} = -4\mathbf{i} + \mathbf{j}$

(iv) $\mathbf{u} = -3\mathbf{i} - 6\mathbf{j}$.

2 Find the magnitude and direction of:

(i) $\begin{pmatrix} 6 \\ -8 \end{pmatrix}$ (ii) $\begin{pmatrix} -4 \\ 0 \end{pmatrix}$ (iii) $\begin{pmatrix} -1 \\ -2 \end{pmatrix}$.

3 Write the resultant, $\mathbf{F}_1 + \mathbf{F}_2$, of the two forces $\mathbf{F}_1 = 10\mathbf{i} + 40\mathbf{j}$ and $\mathbf{F}_2 = 20\mathbf{i} - 10\mathbf{j}$ in component notation and then find its magnitude and direction.

4 Write the resultant of the three forces $\mathbf{F}_1 = -\mathbf{i} + 5\mathbf{j}$, $\mathbf{F}_2 = 2\mathbf{i} - 10\mathbf{j}$ and $\mathbf{F}_3 = -2\mathbf{i} + 7\mathbf{j}$ in component notation and then find its magnitude and direction.

5 (i) Show that $\frac{3}{5}\mathbf{i} + \frac{4}{5}\mathbf{j}$ is a unit vector.

(ii) Find unit vectors in the directions of

(a) $8\mathbf{i} + 6\mathbf{j}$ (b) $\mathbf{i} - \mathbf{j}$.

6 (i) Show that $\frac{1}{3}\mathbf{i} - \frac{2}{3}\mathbf{j} + \frac{2}{3}\mathbf{k}$ is a unit vector.

(ii) Find unit vectors in the directions of

(a) $2\mathbf{i} - 6\mathbf{j} + 3\mathbf{k}$ (b) $\mathbf{i} + \mathbf{j} + \mathbf{k}$.

7 (i) Find a unit vector in the direction of $5\mathbf{i} - 12\mathbf{j}$.

A force $\mathbf{F}$ acts in the direction $5\mathbf{i} - 12\mathbf{j}$ and has magnitude 39 N.

(ii) Use your answer to part (i) to write $\mathbf{F}$ in component form.

8 Write a force **P** of 50 N acting in the direction $7\mathbf{i} + 24\mathbf{j}$ in component form.

9 The position vectors of the points A, B and C are $\mathbf{a} = \mathbf{i} + \mathbf{j} - 2\mathbf{k}$,
$\mathbf{b} = 6\mathbf{i} - 3\mathbf{j} + \mathbf{k}$ and $\mathbf{c} = -2\mathbf{i} + 2\mathbf{j}$.
(i) Find the vectors $\overrightarrow{AC}$, $\overrightarrow{AB}$ and $\overrightarrow{BC}$.
(ii) Find $|\mathbf{a}|$, $|\mathbf{b}|$ and $|\mathbf{c}|$.
(iii) Show that $|\mathbf{a} + \mathbf{b} - \mathbf{c}|$ is *not* equal to $|\mathbf{a}| + |\mathbf{b}| - |\mathbf{c}|$.

Resolving vectors

A vector has magnitude 10 units and it makes an angle of 60° with the **i** direction. How can it be represented in component form?

In the diagram:

$$\frac{AC}{AB} = \cos 60° \qquad \text{and} \qquad \frac{BC}{AB} = \sin 60°$$

$$AC = AB \cos 60° \qquad\qquad BC = AB \sin 60°$$

$$= 10 \cos 60° \qquad\qquad\quad = 10 \sin 60°$$

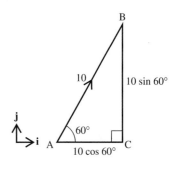

Figure 5.24

The vector can then be written as $10 \cos 60° \, \mathbf{i} + 10 \sin 60° \, \mathbf{j} = 5\mathbf{i} + 8.66\mathbf{j}$ (to 3 sf).

In a similar way, any vector **a** with magnitude a which makes an angle α with the **i** direction can be written in component form as

$$\mathbf{a} = a \cos \alpha \, \mathbf{i} + a \sin \alpha \, \mathbf{j}.$$

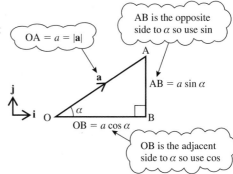

AB is the opposite side to α so use sin

$OA = a = |\mathbf{a}|$

$AB = a \sin \alpha$

$OB = a \cos \alpha$

OB is the adjacent side to α so use cos

Figure 5.25

When α is an obtuse angle, this expression is still true. For example, when $\alpha = 120°$ and $a = 10$,

$$\mathbf{a} = a \cos \alpha \, \mathbf{i} + a \sin \alpha \, \mathbf{j}$$
$$= 10 \cos 120° \, \mathbf{i} + 10 \sin 120° \, \mathbf{j}$$
$$= -5\mathbf{i} + 8.66\mathbf{j}$$

However, it is usually easier to write

$$\mathbf{a} = -10 \cos 60° \, \mathbf{i} + 10 \sin 60° \, \mathbf{j}$$

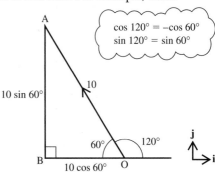

$\cos 120° = -\cos 60°$
$\sin 120° = \sin 60°$

Figure 5.26

EXAMPLE 5.7

Two forces **P** and **Q** have magnitudes 4 and 5 in the directions shown in the diagram.

Find the magnitude and direction of the resultant force **P** + **Q**.

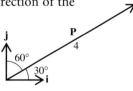
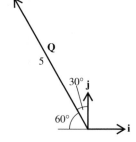

Figure 5.27

SOLUTION

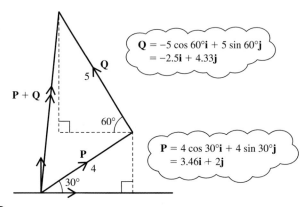

$$\mathbf{Q} = -5\cos 60°\mathbf{i} + 5\sin 60°\mathbf{j}$$
$$= -2.5\mathbf{i} + 4.33\mathbf{j}$$

$$\mathbf{P} = 4\cos 30°\mathbf{i} + 4\sin 30°\mathbf{j}$$
$$= 3.46\mathbf{i} + 2\mathbf{j}$$

Figure 5.28

$$\mathbf{P} + \mathbf{Q} = (3.46\mathbf{i} + 2\mathbf{j}) + (-2.5\mathbf{i} + 4.33\mathbf{j})$$
$$= 0.96\mathbf{i} + 6.33\mathbf{j}$$

This resultant is shown in the diagram on the right.

Magnitude $\quad |\mathbf{P} + \mathbf{Q}| = \sqrt{0.96^2 + 6.33^2}$
$$= \sqrt{40.99}$$
$$= 6.4$$

Direction $\quad \tan\theta = \dfrac{6.33}{0.96}$
$$= 6.59$$
$$\theta = 81.4°$$

The force **P** + **Q** has magnitude 6.4 and direction 81.4° relative to the positive x direction.

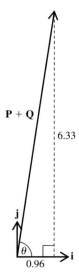

Figure 5.29

Note

If you choose to use the angle that **P** makes with the y axis, 60°, as in figure 5.30, the components are $4\sin 60°\,\mathbf{i} + 4\cos 60°\,\mathbf{j}$.

This is the same as before because $\sin 60° = \cos 30°$ and $\cos 60° = \sin 30°$.

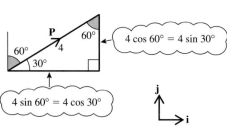

$4\cos 60° = 4\sin 30°$

$4\sin 60° = 4\cos 30°$

Figure 5.30

1 Write down the following vectors in component form in terms of **i** and **j** and in column vector form.

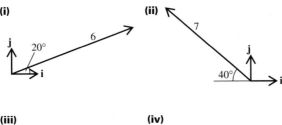

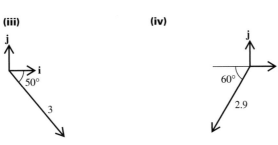

(iii)   **(iv)**

2 Draw a diagram showing each of the following displacements. Write each in component form using unit vectors **i** and **j** in directions east and north respectively.

   (i)   130 km, bearing 060°
   (ii)   250 km, bearing 130°
   (iii) 400 km, bearing 210°
   (iv)   50 miles, bearing 300°

3 A boat has a speed of 4 km h^{-1} in still water and sets its course north-east in an easterly current of 3 km h^{-1}. Write each velocity in component form using unit vectors **i** and **j** in directions east and north and hence find the magnitude and direction of the resultant velocity.

4 A boy walks 30 m north and then 50 m south-west.
   (i)   Draw a diagram to show the boy's path.
   (ii) Write each displacement using column vectors in directions east and north.
   (iii) In which direction should he walk to get directly back to his starting point?

5 **(a)** Write down each of the following vectors in terms of **i** and **j**.
   (b) Find the resultant of each set of vectors in terms of **i** and **j**.

   (i)   **(ii)**

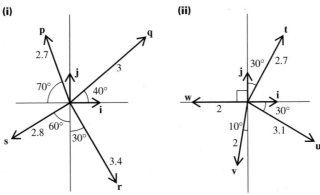

6 (i) Find the distance and bearing of Sean relative to his starting point if he goes for a walk with the following three stages.

Stage 1: 600 m on a bearing 030°

Stage 2: 1 km on a bearing 100°

Stage 3: 700 m on a bearing 340°

(ii) Shona sets off from the same place at the same time as Sean. She walks at the same speed but takes the stages in the order 3–1–2.

How far apart are Sean and Shona at the end of their walks?

7 The diagram shows the journey of a yacht.

Express $\overrightarrow{OA}$, $\overrightarrow{AB}$ and $\overrightarrow{OB}$ as vectors in terms of **i** and **j**, which are unit vectors east and north respectively.

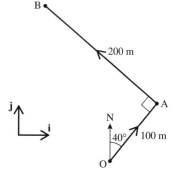

8 A plane is travelling from Plymouth to London at 200 km h^{-1} on a bearing 075°. Due to fog the plane changes direction to fly to Birmingham on a bearing of 015°.

(i) Show the velocity vectors and the change in velocity on a diagram.

(ii) Write down the components of the change in velocity in terms of unit vectors in directions east and north.

9 The diagram shows the big wheel ride at a fairground. The radius of the wheel is 5 m and the length of the arms that support each carriage is 1 m.

Express the position vector of the carriages A, B, C and D in terms of **i** and **j**.

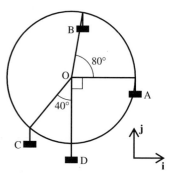

10 A plane completes a journey in three stages. The displacements at each stage, in kilometres, are $3000\mathbf{i} + 4000\mathbf{j}$, $1000\mathbf{i} + 500\mathbf{j}$ and $300\mathbf{i} - 1000\mathbf{j}$, where **i** and **j** are unit vectors in directions east and north respectively.

Express the total journey as a distance and a bearing.

11 Two walkers set off from the same place in different directions. After a period they stop. Their displacements are $\binom{2}{5}$ and $\binom{-3}{4}$ where the distances are in kilometres and the directions are east and north. On what bearing and for what distance does the second walker have to walk to be reunited with the first (who does not move)?

12 Three vectors, **a**, **b**, and **c** are represented by the sides of a triangle ABC as shown in the diagram.

The angle C is θ and $|\mathbf{a}|$, $|\mathbf{b}|$ and $|\mathbf{c}|$ are a, b, and c. Answer each part in terms of θ, a, b, and c.

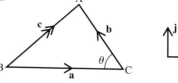

(i) Write **a** and **b** in terms of **i** and **j**.

(ii) Find **a** + **b** and hence $|\mathbf{a} + \mathbf{b}|^2$.

(iii) Use your answer to part (ii) to express c^2 in terms of a, b and θ.

Velocity triangles

When a boat moves in a current or a plane in a wind, the *velocity relative to the water or the air* is the same as the velocity in still water or in still air. It is the velocity the boat or plane would have if there were no current or wind. The resultant vector is found by adding the intended velocity and the velocity of the current or wind.

EXAMPLE 5.8

A swimmer is attempting to cross a river which has a current of $5\,\text{km}\,\text{h}^{-1}$ parallel to its banks. She aims at a point directly opposite her starting point and can swim at $4\,\text{km}\,\text{h}^{-1}$ in still water. Find her resultant velocity.

SOLUTION

The swimmer's velocity in still water and the velocity of the current are shown in the velocity diagram.

By Pythagoras' theorem in triangle ABC:

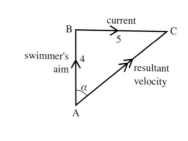

Actual speed $= \sqrt{4^2 + 5^2}$
$= 6.4\,\text{km}\,\text{h}^{-1}$

Direction $\tan \alpha = \dfrac{5}{4}$

$\alpha = 51°$

Figure 5.31

The swimmer has a velocity of $6.4\,\text{km}\,\text{h}^{-1}$ at an angle of $90° - 51° = 39°$ to the bank.

EXAMPLE 5.9

A small motorboat moving at $8\,\text{km}\,\text{h}^{-1}$ relative to the water travels directly between two lighthouses which are 10 km apart, the bearing of the second lighthouse from the first being 135°. The current has a constant speed of $4\,\text{km}\,\text{h}^{-1}$ from the east. Find

(ii) the course that the boat must set

(ii) the time for the journey.

SOLUTION

First draw a clear diagram showing the triangle of velocities. The boat is required to achieve a course of 135° and this must be the direction of the resultant of the boat's own velocity and that of the current.

Draw a line representing the resultant direction and mark on it a point A. Then draw a line AB to represent the current (4 km h^{-1} from the east). Finally, draw a line BC to represent the boat's velocity relative to the water

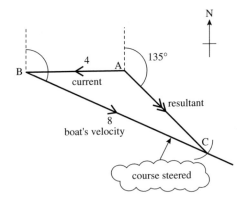

Figure 5.32

(8 km h^{-1}); this must be to the same scale as AB and C must lie on the resultant. (If you were to solve the problem by scale drawing you would need to use a compass to find point C.)

 Notice that the 10 km *displacement* is not part of the *velocity* diagram.

(i) To calculate the course, use the sine rule in triangle ABC.

$\angle$BAC = 135°, AB = 4, BC = 8.

$$\frac{\sin C}{4} = \frac{\sin 135°}{8}$$

$\angle$ACB = 20.7°

$\angle$ABC = 180° − 135° − 20.7° = 24.3°

Therefore the course steered is 90° + 24° = 114° (to the nearest degree).

(ii) The resultant speed is represented by AC and is also found by the sine rule.

$$\frac{AC}{\sin 24.3°} = \frac{8}{\sin 135°}$$

$$AC = 4.656 \text{ (in km h}^{-1}\text{)}$$

Now you can use the resultant speed to find the journey time.

$$\text{Time} = \frac{10}{4.656} = 2.15 \text{ hours.}$$

EXERCISE 5E

Draw diagrams to help you answer these questions.

1 A yacht is sailing at a speed of 6 knots due east but the current is flowing north-west at a speed of 3 knots. Find the resultant speed of the yacht.

2 A plane flies due north at 300 km h^{-1} but a crosswind blows north-west at 40 km h^{-1}. Find the resultant velocity of the plane.

3 A ship has speed 10 knots in still water. It heads due west in a current of 3 knots from 150°.
  (i) Find the actual speed and course of the ship.
  (ii) If the ship is to travel due west, on what course must the captain steer?

4 A woman can paddle a canoe at $3\,\text{ms}^{-1}$ in still water. She wishes to paddle straight across a river which is 20 m wide and flows at a constant speed of $1.5\,\text{ms}^{-1}$ parallel to the banks.
  (i) At what angle to the bank should she paddle?
  (ii) How long does she take to cross?

5 A small motorboat moving at $8\,\text{km h}^{-1}$ relative to the water travels directly between two lighthouses which are 10 km apart, the bearing of the second lighthouse from the first being 150°. The current has a constant speed of $4\,\text{km h}^{-1}$ from the west. Find the course that the boat must set and the time for the journey.

6 A plane leaves Heathrow airport at noon travelling due east towards Prague, 1200 km away. The speed of the plane in still air is $400\,\text{km h}^{-1}$.
  (i) Initially there is no wind. Estimate the time of arrival of the plane.

  After flying for one hour the plane runs into a storm with strong winds of $75\,\text{km h}^{-1}$ from the south-west.
  (ii) If the pilot fails to adjust the heading of the plane how many kilometres from Prague will the plane be after the estimated journey time?
  (iii) What new course should the pilot set?

7 A ferry travels between ports A and B on two islands, 60 km apart. Port A is due west of port B. There is a current in the sea of constant speed $4\,\text{km h}^{-1}$ from a bearing 260°. The ferry can travel at $20\,\text{km h}^{-1}$ in still water.
  (i) Calculate the bearing on which the ferry must head from each port.
  (ii) On a particular journey from A to B, the engines fail when the ferry is half way between the ports. If the coastline near to port B runs north–south, at what distance from port B does the ferry hit land?

8 A gun fires bullets at a speed of $100\,\text{ms}^{-1}$ from a plane moving at $50\,\text{ms}^{-1}$. The bullets can be fired in any direction: what range of speeds do they have?

9 Rain falling vertically hits the windows of an intercity train which is travelling at $50\,\text{ms}^{-1}$. A passenger watches the rain streaks on the window and estimates them to be at an angle of 20° to the horizontal.
  (i) Calculate the speed the rain would run down a stationary window.
  (ii) The train starts to slow down with a uniform deceleration and 1.5 minutes later the streaks are at 45° to the horizontal. What is the deceleration of the train?

INVESTIGATION

Investigate how it is possible to sail into the wind.

1 A vector has both magnitude and direction.

2 Vectors may be represented in either magnitude–direction form or in component form.

Magnitude–direction form   **Component form**

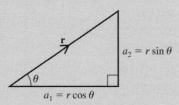   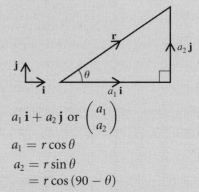

Magnitude, r direction, θ $a_1\mathbf{i} + a_2\mathbf{j}$ or $\begin{pmatrix} a_1 \\ a_2 \end{pmatrix}$

Where $\quad r = \sqrt{a_1{}^2 + a_2{}^2}$ $a_1 = r\cos\theta$

and $\quad \tan\theta = \dfrac{a_2}{a_1}$ $a_2 = r\sin\theta$

$\qquad\qquad\qquad\qquad = r\cos(90 - \theta)$

3 When two or more vectors are added, the *resultant* is obtained. Vector addition may be done graphically or algebraically.

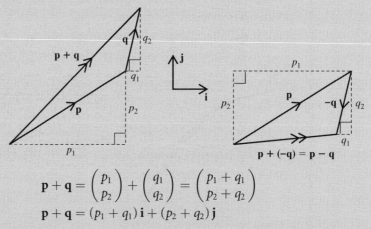

$$\mathbf{p} + \mathbf{q} = \begin{pmatrix} p_1 \\ p_2 \end{pmatrix} + \begin{pmatrix} q_1 \\ q_2 \end{pmatrix} = \begin{pmatrix} p_1 + q_1 \\ p_2 + q_2 \end{pmatrix}$$

$$\mathbf{p} + \mathbf{q} = (p_1 + q_1)\mathbf{i} + (p_2 + q_2)\mathbf{j}$$

4 Multiplication by a scalar

$$n(a_1\mathbf{i} + a_2\mathbf{j}) = na_1\mathbf{i} + na_2\mathbf{j} \qquad n\begin{pmatrix} a_1 \\ a_2 \end{pmatrix} = \begin{pmatrix} na_1 \\ na_2 \end{pmatrix}$$

5 The *position vector* of a point P is $\overrightarrow{OP}$, its displacement from a fixed origin.

6 When A and B have position vectors $\mathbf{a}$ and $\mathbf{b}$, $\overrightarrow{AB} = \mathbf{b} - \mathbf{a}$.

7 *Equal vectors* have equal magnitude and are in the same direction.
$$p_1\mathbf{i} + p_2\mathbf{j} = q_1\mathbf{i} + q_2\mathbf{j} \Rightarrow p_1 = q_1 \text{ and } p_2 = q_2$$

8 When $p_1\mathbf{i} + p_2\mathbf{j}$ and $q_1\mathbf{i} + q_2\mathbf{j}$ are parallel, $\frac{p_1}{q_1} = \frac{p_2}{q_2}$.

9 The *unit vector* in the direction of $a_1\mathbf{i} + a_2\mathbf{j}$ is $\dfrac{a_1}{\sqrt{a_1{}^2 + a_2{}^2}}\mathbf{i} + \dfrac{a_2}{\sqrt{a_2{}^2 + a_2{}^2}}\mathbf{j}$.

6

Projectiles

Swift of foot was Hiawatha;
He could shoot an arrow from him,
And run forward with such fleetness,
That the arrow fell behind him!
Strong of arm was Hiawatha;
He could shoot ten arrows upwards,
Shoot them with such strength and swiftness,
That the last had left the bowstring,
Ere the first to earth had fallen!

The Song of Hiawatha, Longfellow

Look at the water jet in the picture. Every drop of water in a water jet follows its own path which is called its *trajectory*. You can see the same sort of trajectory if you throw a small object across a room. Its path is a parabola. Objects moving through the air like this are called projectiles.

Modelling assumptions for projectile motion

The path of a cricket ball looks parabolic, but what about a boomerang? There are modelling assumptions which must be satisfied for the motion to be parabolic. These are

- a projectile is a particle
- it is not powered
- the air has no effect on its motion.

Equations for projectile motion

A projectile moves in two dimensions under the action of only one force, the force of gravity, which is constant and acts vertically downwards. This means that the acceleration of the projectile is g ms^{-2} vertically downwards and there is no horizontal acceleration. You can treat the horizontal and vertical motion separately using the equations for constant acceleration.

To illustrate the ideas involved, think of a ball being projected with a speed of 20 ms^{-1} at $60°$ to the ground as illustrated in figure 6.1. This could be a first model for a football, a chip shot from the rough at golf or a lofted shot at cricket.

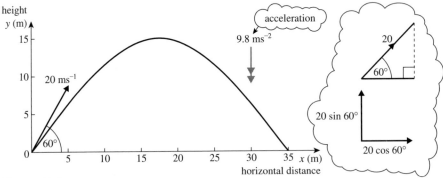

Figure 6.1

Using axes as shown, the components are:

	Horizontal	Vertical
Initial position	0	0
Acceleration	$a_x = 0$	$a_y = -9.8$

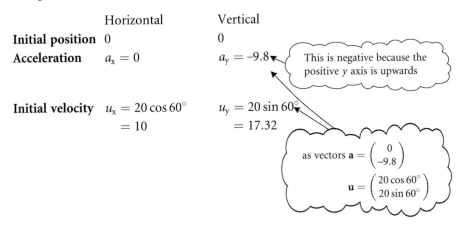

This is negative because the positive y axis is upwards

	Horizontal	Vertical
Initial velocity	$u_x = 20 \cos 60°$	$u_y = 20 \sin 60°$
	$= 10$	$= 17.32$

as vectors $\mathbf{a} = \begin{pmatrix} 0 \\ -9.8 \end{pmatrix}$

$\mathbf{u} = \begin{pmatrix} 20 \cos 60° \\ 20 \sin 60° \end{pmatrix}$

Using $v = u + at$ in the two directions gives the components of velocity.

Velocity   Horizontal   Vertical

$a_x = 0 \Rightarrow v_x$ is constant

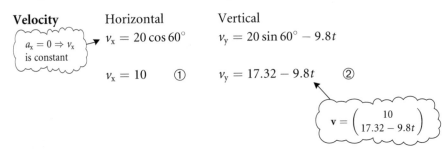

$$v_x = 20 \cos 60° \qquad v_y = 20 \sin 60° - 9.8t$$

$$v_x = 10 \quad ① \qquad v_y = 17.32 - 9.8t \quad ②$$

$\mathbf{v} = \begin{pmatrix} 10 \\ 17.32 - 9.8t \end{pmatrix}$

Using $s = ut + \frac{1}{2}at^2$ in both directions gives the components of position.

Position Horizontal Vertical

$$x = (20 \cos 60°)t \qquad y = (20 \sin 60°)t - 4.9t^2$$
$$x = 10t \qquad ③ \qquad y = 17.32t - 4.9t^2 \qquad ④$$

$$\mathbf{r} = \begin{pmatrix} 10t \\ 17.32t - 4.9t^2 \end{pmatrix}$$

You can summarise these results in a table.

	Horizontal motion		Vertical motion	
initial position	0		0	
a	0		−9.8	
u	$u_x = 20 \cos 60° = 10$		$u_y = 20 \sin 60° = 17.32$	
v	$v_x = 10$	①	$v_y = 17.32 - 9.8t$	②
r	$x = 10t$	③	$y = 17.32t - 4.9t^2$	④

The four equations ①, ②, ③ and ④ for velocity and position can be used to find several things about the motion of the ball.

? What is true at

(i) the top-most point of the path of the ball?

(ii) the point where it is just about to hit the ground?

When you have decided the answer to these questions you have sufficient information to find the greatest height reached by the ball, the time of flight and the total distance travelled horizontally before it hits the ground. This is called the range of the ball.

THE MAXIMUM HEIGHT

When the ball is at its maximum height, H m, the *vertical* component of its velocity is zero. It still has a horizontal component of 10 ms^{-1} which is constant.

Equation ② gives the vertical component as

$$v_y = 17.32 - 9.8t$$
At the top: $0 = 17.32 - 9.8t$
$$t = \frac{17.32}{9.8}$$
$$= 1.767$$

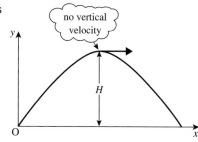

Figure 6.2

To find the maximum height, you now need to find y at this time. Substituting for t in equation ④,

$$y = 17.32t - 4.9t^2$$
$$y = 17.32 \times 1.767 - 4.9 \times 1.767^2$$
$$= 15.3$$

The maximum height is 15.3 m.

THE TIME OF FLIGHT

The flight ends when the ball returns to the ground, that is when $y = 0$. Substituting $y = 0$ in equation ④,

$$y = 17.32t - 4.9t^2$$
$$17.32t - 4.9t^2 = 0$$
$$t(4.9t - 17.3) = 0$$
$$t = 0 \text{ or } t = 3.53$$

Clearly $t = 0$ is the time when the ball is thrown, so $t = 3.53$ is the time when it lands and the flight time is 3.53 s.

THE RANGE

The range, R m, of the ball is the horizontal distance it travels before landing.

R is the value of x when $y = 0$.

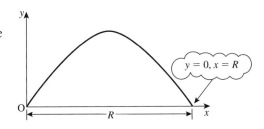

$y = 0, x = R$

Figure 6.3

R can be found by substituting $t = 3.53$ in equation ③: $x = 10t$. The range is $10 \times 3.53 = 35.3$ m.

? **1** Notice in this example that the time to maximum height is half the flight time. Is this always the case?

2 Decide which of the following could be modelled as projectiles.

a balloon	a bird	a bullet shot from a gun	a glider
a golf ball	a parachutist	a rocket	a tennis ball

What special conditions would have to apply in particular cases?

In this exercise take upwards as positive and use $9.8\,ms^{-2}$ for g unless otherwise stated. All the projectiles start at the origin.

1 In each case you are given the initial velocity of a projectile.
  (a) Draw a diagram showing the initial velocity and path.
  (b) Write down the horizontal and vertical components of the initial velocity.
  (c) Write down equations for the velocity after time t seconds.
  (d) Write down equations for the position after time t seconds.
      (i) $10\,ms^{-1}$ at $35°$ above the horizontal.
      (ii) $2\,ms^{-1}$ horizontally, $5\,ms^{-1}$ vertically.
      (iii) $4\,ms^{-1}$ horizontally.
      (iv) $10\,ms^{-1}$ at $13°$ below the horizontal.
      (v) $U\,ms^{-1}$ at angle α above the horizontal.
      (vi) $u_0\,ms^{-1}$ horizontally, $v_0\,ms^{-1}$ vertically.

2 In each case find
  (a) the time taken for the projectile to reach its highest point
  (b) the maximum height.
      (i) Initial velocity $5\,ms^{-1}$ horizontally and $14.7\,ms^{-1}$ vertically.
      (ii) Initial velocity $10\,ms^{-1}$ at $30°$ above the horizontal. Use $10\,ms^{-2}$ for g.

3 In each case find
  (a) the time of flight of the projectile
  (b) the horizontal range.
      (i) Initial velocity $20\,ms^{-1}$ horizontally and $19.6\,ms^{-1}$ vertically.
      (ii) Initial velocity $5\,ms^{-1}$ at $60°$ above the horizontal.

Projectile problems

When doing projectile problems, you can treat each direction separately or you can write them both together as vectors. Example 6.1 shows both methods.

EXAMPLE 6.1

A ball is thrown horizontally at $5\,ms^{-1}$ out of a window $4\,m$ above the ground.

(i) How long does it take to reach the ground?
(ii) How far from the building does it land?
(iii) What is its speed just before it lands and at what angle to the ground is it moving?

SOLUTION

The diagram shows the path of the ball. It is important to decide at the outset where the origin and axes are. You may choose any axes that are suitable, but you must specify them carefully to avoid making mistakes. Here the origin is taken to be at ground level below the point of projection of the ball and upwards is positive. With these axes, the acceleration is $-g\,ms^{-2}$.

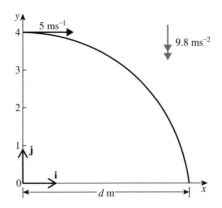

Figure 6.4

Method 1:

Resolving into components

Position: Using axes as shown and $s = s_0 + ut + \frac{1}{2}at^2$ in the two directions

Horizontally $x_0 = 0,\ u_x = 5,\ a_x = 0$

                $x = 5t$ ①

Vertically $y_0 = 4,\ u_y = 0,\ a_y = -9.8$

                $y = 4 - 4.9t^2$ ②

(i) The ball reaches the ground when $y = 0$. Substituting in equation ② gives

$$0 = 4 - 4.9t^2$$
$$t^2 = \frac{4}{4.9}$$
$$t = 0.904$$

The ball hits the ground after 0.9 s.

(ii) When the ball lands $x = d$ so, from equation ①,

$d = 5t = 5 \times 0.904 = 4.52$. The ball lands 4.52 m from the building.

Velocity: Using $v = u + at$ in the two directions

Horizontally $v_x = 5 + 0$

Vertically $v_y = 0 - 9.8t$

(iii) To find the speed and direction just before it lands:

The ball lands when $t = 0.904$ so $v_x = 5$ and $v_y = -8.86$.

The components of velocity are shown in the diagram.
The speed of the ball is $\sqrt{5^2 + 8.86^2} = 10.17\ \text{ms}^{-1}$.
It hits the ground moving downwards at an angle α to
the horizontal where

$$\tan \alpha = \frac{8.86}{5}$$
$$\alpha = 60.6°$$

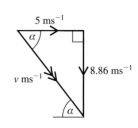

Figure 6.5

Method 2:

Using vectors

Using the unit vectors shown, the initial position is $\mathbf{r}_0 = 4\mathbf{j}$ and the ball hits the ground when $\mathbf{r} = d\mathbf{i}$. The initial velocity, $\mathbf{u} = 5\mathbf{i}$ and the acceleration $\mathbf{a} = -9.8\mathbf{j}$. In column vectors these are

$$\mathbf{r}_0 = \begin{pmatrix} 0 \\ 4 \end{pmatrix} \quad \mathbf{r} = \begin{pmatrix} d \\ 0 \end{pmatrix} \quad \mathbf{u} = \begin{pmatrix} 5 \\ 0 \end{pmatrix} \quad \mathbf{a} = \begin{pmatrix} 0 \\ -9.8 \end{pmatrix}$$

Using

$$\mathbf{r} = \mathbf{r}_0 + \mathbf{u}t + \tfrac{1}{2}\mathbf{a}t^2$$

$$\begin{pmatrix} d \\ 0 \end{pmatrix} = \begin{pmatrix} 0 \\ 4 \end{pmatrix} + \begin{pmatrix} 5 \\ 0 \end{pmatrix}t + \tfrac{1}{2}\begin{pmatrix} 0 \\ -9.8 \end{pmatrix}t^2$$

$$d = 5t \qquad\qquad\qquad\qquad ①$$

and

$$0 = 4 - 4.9t^2 \qquad\qquad ②$$

(i) Equation ② gives $t = 0.904$ and substituting this into ① gives **(ii)** $d = 4.52$.

(iii) The speed and direction of motion are the magnitude and direction of the velocity of the ball. Using

$$\mathbf{v} = \mathbf{u} + \mathbf{a}t$$

$$\begin{pmatrix} v_x \\ v_y \end{pmatrix} = \begin{pmatrix} 5 \\ 0 \end{pmatrix} + \begin{pmatrix} 0 \\ -9.8 \end{pmatrix}t$$

So when $t = 0.904$, $\begin{pmatrix} v_x \\ v_y \end{pmatrix} = \begin{pmatrix} 5 \\ -8.86 \end{pmatrix}$

You can find the speed and angle as before.

Notice that in both methods the time forms a link between the motions in the two directions. You can often find the time from one equation and then substitute it in another to find out more information.

Representing projectile motion by vectors

The diagram shows a possible path for a marble which is thrown across a room from the moment it leaves the hand until just before it hits the floor.

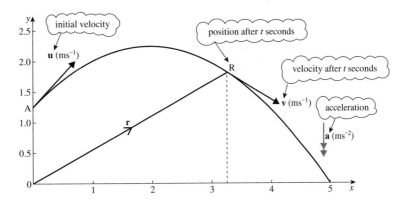

Figure 6.6

The vector $\mathbf{r} = \overrightarrow{OR}$ is the position vector of the marble after a time t seconds and the vector $\mathbf{v}$ represents its velocity in ms^{-1} at that instant of time (to a different scale).

 Notice that the graph shows the trajectory of the marble. It is its path through space, not a position–time graph.

You can use equations for constant acceleration in vector form to describe the motion as in Example 6.1, Method 2.

velocity $\qquad\qquad \mathbf{v} = \mathbf{u} + \mathbf{a}t$

displacement $\qquad \mathbf{r} - \mathbf{r}_0 = \mathbf{u}t + \frac{1}{2}\mathbf{a}t^2$ so $\mathbf{r} = \mathbf{r}_0 + \mathbf{u}t + \frac{1}{2}\mathbf{a}t^2$

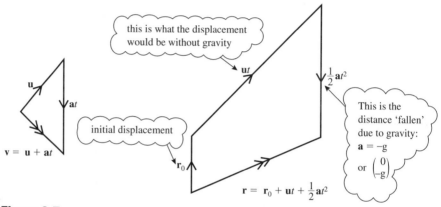

Figure 6.7

 Always check whether or not the projectile starts at the origin. The change in position is the vector $\mathbf{r} - \mathbf{r}_0$. This is the equivalent of $s - s_0$ in one dimension.

EXERCISE 6B

In this exercise take upwards as positive and use 9.8 ms^{-2} for g in numerical questions.

1 In each case
   (a) draw a diagram showing the initial velocity and path and write in vector form
   (b) the velocity
   (c) the position after time t s.
      (i) Initial position (0, 10 m); initial velocity 4 ms^{-1} horizontally.
      (ii) Initial position (0, 7 m); initial velocity 10 ms^{-1} at 35° above the horizontal.
      (iii) Initial position (0, 20 m); initial velocity 10 ms^{-1} at 13° below the horizontal.
      (iv) Initial position O; initial velocity $\begin{pmatrix} 7 \\ 24 \end{pmatrix}$ ms^{-1}.
      (v) Initial position (a, b) m; initial velocity $\begin{pmatrix} u_0 \\ v_0 \end{pmatrix}$ ms^{-1}.

2 In each case find

(a) the time taken for the projectile to reach its highest point

(b) the maximum height above the origin.

(i) Initial position (0, 15 m) velocity 5 ms⁻¹ horizontally and 14.7 ms⁻¹ vertically.

(ii) Initial position (0, 10 m); initial velocity $\begin{pmatrix} 5 \\ 3 \end{pmatrix}$ ms⁻¹.

3 Find the horizontal range for these projectiles which start from the origin.

(i) Initial velocity $\begin{pmatrix} 2 \\ 7 \end{pmatrix}$ ms⁻¹.

(ii) Initial velocity $\begin{pmatrix} 7 \\ 2 \end{pmatrix}$ ms⁻¹.

(iii) Sketch the paths of these two projectile using the same axes.

Further examples

EXAMPLE 6.2

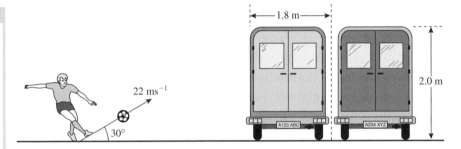

Figure 6.8

In this question use 10 ms⁻² for g and neglect air resistance.

In an attempt to raise money for a charity, participants are sponsored to kick a ball over some vans. The vans are each 2 m high and 1.8 m wide and stand on horizontal ground. One participant kicks the ball at an initial speed of 22 ms⁻¹ inclined at 30° to the horizontal.

(i) What are the initial values of the vertical and horizontal components of velocity?

(ii) Show that while in flight the vertical height y metres at time t seconds satisfies the equation $y = 11t - 5t^2$.

Calculate at what times the ball is at least 2 m above the ground.

The ball should pass over as many vans as possible.

(iii) Deduce that the ball should be placed about 3.8 m from the first van and find how many vans the ball will clear.

(iv) What is the greatest vertical distance between the ball and the top of the vans?

[MEI]

SOLUTION

(i) *Initial velocity*

horizontally: $22 \cos 30° = 19.05 \text{ ms}^{-1}$

vertically: $22 \sin 30° = 11 \text{ ms}^{-1}$

22
30°
22 sin 30°
22 cos 30°

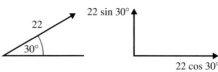

Figure 6.9

(ii) *When the ball is above 2 m*

Using axes as shown and $s = ut + \frac{1}{2}at^2$ vertically

$$\Rightarrow \quad y = 11t - 5t^2$$

The ball is 2 m above the ground when $y = 2$, then

y
2
0
x

Figure 6.10

$$2 = 11t - 5t^2$$
$$5t^2 - 11t + 2 = 0$$
$$(5t - 1)(t - 2) = 0$$
$$t = 0.2 \text{ or } 2$$

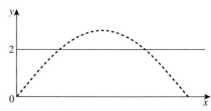

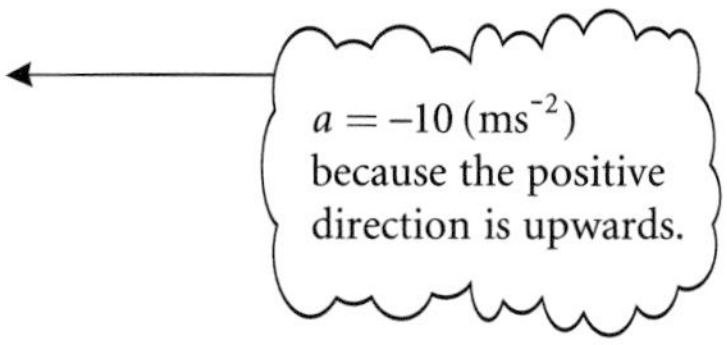

a = −10 (ms⁻²) because the positive direction is upwards.

The ball is at least 2 m above the ground when $0.2 \leqslant t \leqslant 2$

(iii) *How many vans?*

Horizontally, $s = ut + \frac{1}{2}at^2$ with $a = 0$

$$\Rightarrow \quad x = 19.05t$$

When $t = 0.2$, $\quad x = 3.81$ (at A)

when $t = 2$, $\quad x = 38.1$ (at B)

To clear as many vans as possible, the ball should be placed about 3.8 m in front of the first van.

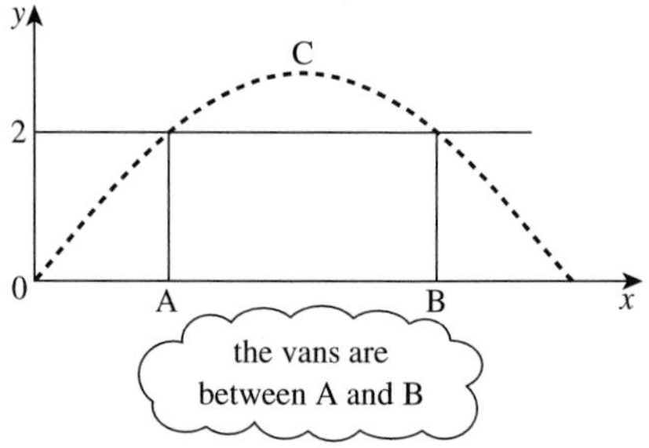

y
C
2
0
A
B
x
the vans are between A and B

Figure 6.11

$$AB = 38.1 - 3.81 \text{ m} = 34.29 \text{ m}$$

$$\frac{34.29}{1.8} = 19.05.$$

The maximum possible number of vans is 19.

(iv) *Maximum height*

At the top (C), vertical velocity = 0, so using $v = u + at$ vertically

$$\Rightarrow \quad 0 = 11 - 10t$$
$$t = 1.1.$$

Substituting in $y = 11t - 5t^2$, maximum height is

$11 \times 1.1 - 5 \times 1.1^2 = 6.05 \text{ m}$

The ball clears the tops of the vans by about 4 m.

EXAMPLE 6.3

Sharon is diving into a swimming pool. During her flight she may be modelled as a particle. Her initial velocity is 1.8 ms^{-1} at angle 30° above the horizontal and initial position 3.1 m above the water. Air resistance may be neglected.

(i)   Find the greatest height above the water that Sharon reaches during her dive.

(ii) Show that the time t, in seconds, that it takes Sharon to reach the water is given by $4.9t^2 - 0.9t - 3.1 = 0$ and solve this equation to find t.
       Explain the significance of the other root to the equation.

Just as Sharon is diving a small boy jumps into the swimming pool. He hits the water at a point in line with the diving board and 1.5 m from its end.

(iii) Is there an accident?

SOLUTION

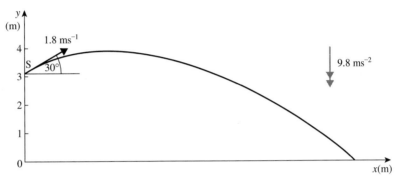

Figure 6.12

Referring to the axes shown

	Horizontal motion		Vertical motion	
initial position	0		3.1	
a	0		–9.8	
u	$u_x = 1.8 \cos 30° = 1.56$		$u_y = 1.8 \sin 30° = 0.9$	
v	$v_x = 1.56$	①	$v_y = 0.9 - 9.8t$	②
r	$x = 1.56t$	③	$y = 3.1 + 0.9t - 4.9t^2$	④

(i)   At the top $v_y = 0$ $0 = 0.9 - 9.8t$ from ②
                                    $t = 0.092$

       When $t = 0.092$ $y = 3.1 + 0.9 \times 0.092 - 4.9 \times 0.092^2$ from ④
                                    $= 3.14$

       Sharon's greatest height above the water is 3.14 m.

(ii) Sharon reaches the water when $y = 0$

$$0 = 3.1 + 0.9t - 4.9t^2 \qquad \text{from ④}$$
$$4.9t^2 - 0.9t - 3.1 = 0$$

$$t = \frac{0.9 \pm \sqrt{0.9^2 + 4 \times 4.9 \times 3.1}}{9.8}$$

$$t = -0.71 \text{ or } 0.89$$

Sharon hits the water after 0.89 s. The negative value of t gives the point on the parabola at water level to the left of the point (S) where Sharon dives.

(iii) At time t the horizontal distance from diving board,

$$x = 1.56t \qquad\qquad \text{from } ③$$

When Sharon hits the water $\qquad x = 1.56 \times 0.89 = 1.39$

Assuming that the particles representing Sharon and the boy are located at their centres of mass, the difference of 11 cm between 1.39 m and 1.5 m is not sufficient to prevent an accident.

Note

When the point S is taken as the origin in the above example, the initial position is $(0, 0)$ and $y = 0.9t - 4.9t^2$. In this case, Sharon hits the water when $y = -3.1$. This gives the same equation for t.

EXAMPLE 6.4

A boy kicks a small ball from the floor of a gymnasium with an initial velocity of 12 ms^{-1} inclined at an angle α to the horizontal. Air resistance may be neglected.

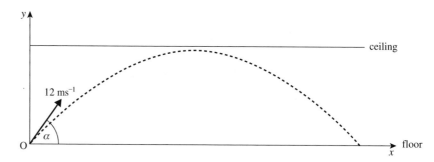

Figure 6.13

(i) Write down expressions in terms of α for the vertical speed of the ball and vertical height of the ball after t seconds.

The ball just fails to touch the ceiling which is 4 m high. The highest point of the motion of the ball is reached after T seconds.

(ii) Use one of your expressions to show that $6 \sin \alpha = 5T$ and the other to form a second equation involving $\sin \alpha$ and T.

(iii) Eliminate $\sin \alpha$ from your two equations to show that T has a value of about 0.89.

(iv) Find the horizontal range of the ball when kicked at 12 ms^{-1} from the floor of the gymnasium so that it just misses the ceiling.

Use 10 ms^{-2} for g in this question.

[MEI]

SOLUTION

(i) *Vertical components*

speed $\quad v_y = 12 \sin \alpha - 10t \quad$ ①

height $\quad y = (12 \sin \alpha)t - 5t^2 \quad$ ②

acceleration (ms^{-2})   initial velocity (ms^{-1})

Figure 6-14

(ii) *Time to highest point*

At the top $v_y = 0$ and $t = T$, so equation ① gives

$$12 \sin \alpha - 10T = 0$$
$$12 \sin \alpha = 10T$$
$$6 \sin \alpha = 5T \qquad\qquad ③$$

When $t = T$, $y = 4$ so from ②

$$4 = 12 \sin \alpha \, T - 5T^2 \qquad\qquad ④$$

(iii) Substituting for $6 \sin \alpha$ from ③ into ④ gives

$$4 = 2 \times 5T \times T - 5T^2$$
$$4 = 5T^2$$
$$T = \sqrt{0.8} = 0.89$$

(iv) *Range*

The path is symmetrical so the time of flight is $2T$ seconds.

Horizontally $a = 0$ and $u_x = 12 \cos \alpha$

$$\Rightarrow \quad x = (12 \cos \alpha) \, t$$

The range is $12 \cos \alpha \times 2T = 21.36 \cos \alpha$ m.

From ③ $\quad 6 \sin \alpha = 5T = 4.45$

$$\alpha = 47.87°$$

The range is $21.47 \cos 47.87° = 14.4$ m.

? Two marbles start simultaneously from the same height. One (P) is dropped and the other (Q) is projected horizontally. Which reaches the ground first?

Use $9.8 \, ms^{-2}$ for g in this exercise unless otherwise specified.

1 A ball is thrown from a point at ground level with velocity $20 \, ms^{-1}$ at $30°$ to the horizontal. The ground is level and horizontal and you should ignore air resistance. Take g to be $10 \, ms^{-2}$ and specify suitable axes.
  (i) Find the horizontal and vertical components of the ball's initial velocity.
  (ii) Find the horizontal and vertical components of the ball's acceleration.
  (iii) Find the horizontal distance travelled by the ball before its first bounce.
  (iv) Find how long the ball takes to reach maximum height.
  (v) Find the maximum height reached by the ball.

2 Nick hits a golf ball with initial velocity $50\,\mathrm{ms}^{-1}$ at $35°$ to the horizontal.

  (i) Find the horizontal and vertical components of the ball's initial velocity.

  (ii) Specify suitable axes and calculate the position of the ball at one second intervals for the first six seconds of its flight.

  (iii) Draw a graph of the path of the ball (its trajectory) and use it to estimate

  (a) the maximum height of the ball

  (b) the horizontal distance the ball travels before bouncing.

  (iv) Calculate the maximum height the ball reaches and the horizontal distance it travels before bouncing. Compare your answers with the estimates you found from your graph.

  (v) State the modelling assumptions you made in answering this question.

3 Clare scoops a hockey ball off the ground, giving it an initial velocity of $19\,\mathrm{ms}^{-1}$ at $25°$ to the horizontal.

  (i) Find the horizontal and vertical components of the ball's initial velocity.

  (ii) Find the time that elapses before the ball hits the ground.

  (iii) Find the horizontal distance the ball travels before hitting the ground.

  (iv) Find how long it takes for the ball to reach maximum height.

  (v) Find the maximum height reached.

  A member of the opposing team is standing $20\,\mathrm{m}$ away from Clare in the direction of the ball's flight.

  (vi) How high is the ball when it passes her? Can she stop the ball?

4 A footballer is standing $30\,\mathrm{m}$ in front of the goal. He kicks the ball towards the goal with velocity $18\,\mathrm{ms}^{-1}$ and angle $55°$ to the horizontal. The height of the goal's crossbar is $2.5\,\mathrm{m}$. Air resistance and spin may be ignored.

  (i) Find the horizontal and vertical components of the ball's initial velocity.

  (ii) Find the time it takes for the ball to cross the goal-line.

  (iii) Does the ball bounce in front of the goal, go straight into the goal or go over the crossbar?

  In fact the goalkeeper is standing $5\,\mathrm{m}$ in front of the goal and will stop the ball if its height is less than $2.8\,\mathrm{m}$ when it reaches him.

  (iv) Does the goalkeeper stop the ball?

5 A plane is flying at a speed of $300\,\mathrm{ms}^{-1}$ and maintaining an altitude of $10\,000\,\mathrm{m}$ when a bolt becomes detached. Ignoring air resistance, find

  (i) the time that the bolt takes to reach the ground

  (ii) the horizontal distance between the point where the bolt leaves the plane and the point where it hits the ground

  (iii) the speed of the bolt when it hits the ground

  (iv) the angle to the horizontal at which the bolt hits the ground.

6 Reena is learning to serve in tennis. She hits the ball from a height of 2 m. For her serve to be legal it must pass over the net which is 12 m away from her and 0.91 m high, and it must land within 6.4 m of the net.

Make the following modelling assumptions to answer the questions.

- She hits the ball horizontally.
- Air resistance may be ignored.
- The ball may be treated as a particle.
- The ball does not spin.
- She hits the ball straight down the middle of the court.

(i) How long does the ball take to fall to the level of the top of the net?

(ii) How long does the ball take from being hit to first reaching the ground?

(iii) What is the lowest speed with which Reena must hit the ball to clear the net?

(iv) What is the greatest speed with which she may hit it if it is to land within 6.4 m of the net?

7 A stunt motorcycle rider attempts to jump over a gorge 50 m wide. He uses a ramp at $25°$ to the horizontal for his take-off and has a speed of $30 \, \text{ms}^{-1}$ at this time.

(i) Assuming that air resistance is negligible, find out whether the rider crosses the gorge successfully.

The stunt man actually believes that in any jump the effect of air resistance is to reduce his distance by 40%.

(ii) Calculate his minimum safe take-off speed for this jump.

8 To kick a goal in rugby you must kick the ball over the crossbar of the goal posts (height 3.0 m), between the two uprights. Dafydd Evans attempts a kick from a distance of 35 m. The initial velocity of the ball is $20 \, \text{ms}^{-1}$ at $30°$ to the horizontal. The ball is aimed between the uprights and no spin is applied.

(i) How long does it take the ball to reach the goal posts?

(ii) Does it go over the crossbar?

Later in the game, Dafydd takes another kick from the same position and hits the crossbar.

(iii) Given that the initial velocity of the ball in this kick was also at $30°$ to the horizontal, find the initial speed.

Many rugby kickers choose to give the ball spin.

(iv) What effect does spin have upon the flight of the ball?

9 In this question take g to be $10 \, \text{ms}^{-2}$. A catapult projects a small pellet at speed $20 \, \text{ms}^{-1}$ and can be directed at any angle to the horizontal.

(i) Find the range of the catapult when the angle of projection is
  (a) $30°$ (b) $40°$ (c) $45°$ (d) $50°$ (e) $60°$.

(ii) Show algebraically that the range is the same when the angle of projection is α as it is when the angle is $90 - \alpha$.

The catapult is angled with the intention that the pellet should hit a point on the ground 36 m away.

(iii) Verify that one appropriate angle of projection would be 32.1° and write down another suitable angle.

In fact the angle of projection from the catapult is liable to error.

(iv) Find the distance by which the pellet misses the target in each of the cases in (iii) when the angle of projection is subject to an error of +0.5°. Which angle should you use for greater accuracy?

10 A cricketer hits the ball on the half-volley, that is when the ball is at ground level. The ball leaves the ground at an angle of 30° to the horizontal and travels towards a fielder standing on the boundary 60 m away.

(i) Find the initial speed of the ball if it hits the ground for the first time at the fielder's feet.

(ii) Find the initial speed of the ball if it is at a height of 3.2 m (well outside the fielder's reach) when it passes over the fielder's head.

In fact the fielder is able to catch the ball without moving provided that its height, h m, when it reaches him satisfies the inequality $0.25 \leqslant h \leqslant 2.1$.

(iii) Find a corresponding range of values for u, the initial speed of the ball.

11 A horizontal tunnel has a height of 3 m. A ball is thrown inside the tunnel with an initial speed of 18 ms^{-1}. What is the greatest horizontal distance that the ball can travel before it bounces for the first time?

12

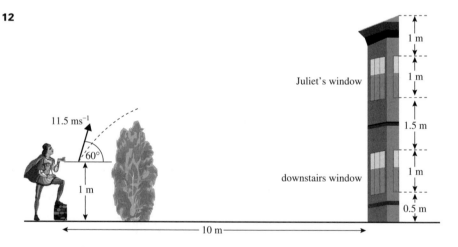

The picture shows Romeo trying to attract Juliet's attention without her nurse, who is in a downstairs room, noticing. He stands 10 m from the house and lobs a small pebble at her bedroom window. Romeo throws the pebble from a height of 1 m with a speed of 11.5 ms^{-1} at an angle of 60° to the horizontal.

(i) How long does the pebble take to reach the house?

(ii) Does the pebble hit Juliet's window, the wall of the house or the downstairs room window?

(iii) What is the speed of the pebble when it hits the house?   [MEI]

13 Use $g = 10\,\text{ms}^{-2}$ in this question.

A firework is buried so that its top is at ground level and it projects sparks all at a speed of $8\,\text{ms}^{-1}$. Air resistance may be neglected.

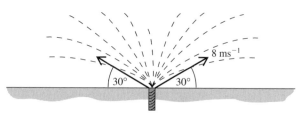

(i) Calculate the height reached by a spark projected vertically and explain why no spark can reach a height greater than this.

(ii) For a spark projected at $30°$ to the horizontal over horizontal ground, show that its height in metres t seconds after projection is $4t - 5t^2$ and hence calculate the distance it lands from the firework.

(iii) For what angle of projection will a spark reach a maximum height of 2 m?

[MEI]

14 A small stone is projected horizontally from a height of 19.6 m over horizontal ground at a speed of $15\,\text{ms}^{-1}$. Air resistance may be neglected.

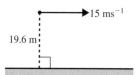

(i) Calculate the time which elapses between projection and the stone hitting the ground and show that the horizontal distance travelled is 30 m.

In a new situation two small stones are projected horizontally towards each other at the same instant over horizontal ground. Initially the stones are 19.6 m above the ground and are 50 m apart. The initial speeds of the stones are $15\,\text{ms}^{-1}$ and $v\,\text{ms}^{-1}$.

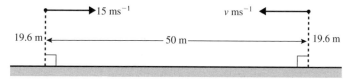

(ii) Explain briefly why the stones will be at the same height as one another above the ground after the elapse of any time interval and hence must collide if they do not hit the ground first.

(iii) Find the value of v if the stones collide 4.9 m above the ground.

(iv) For what values of v will the stones not collide in the air?

[MEI]

15 In this question use $10\,\text{ms}^{-2}$ for g. When attempting to kick a ball from the ground, a boy finds that the angle of projection affects the speed at which he can kick the ball. The horizontal and

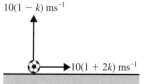

vertical components of projection are $10(1 + 2k)\,\text{ms}^{-1}$ and $10(1 - k)\,\text{ms}^{-1}$ respectively, where k is a number such that $0 \leqslant k \leqslant 1$. The ball is kicked over horizontal ground and air resistance may be neglected.

(i) Find the speed and angle of projection in the cases

    (a) $k = 0$ **(b)** $k = 1$.

(ii) Find expressions in terms of k for the horizontal and vertical positions of the ball t seconds after projection.

(iii) Find, in terms of k, how long the ball is in the air before it hits the ground again.

(iv) Show that the range is $20(1 + k - 2k^2)\,\text{m}$.

Given that $20(1 + k - 2k^2)$ can be written as $\frac{5}{2}\left(9 - 16\left(k - \frac{1}{4}\right)^2\right)$ for all k, write down the value of the maximum range.

[MEI]

The path of a projectile

Look at the equations

$$x = 20t$$
$$y = 6 + 30t - 5t^2$$

They represent the path of a projectile.

? What is the initial velocity of the projectile? What is its initial position? What value of g is assumed?

These equations give x and y in terms of a third variable t. (They are called *parametric equations* and t is the *parameter*.)

You can find the *cartesian equation* connecting x and y directly by eliminating t as follows

$$x = 20t \Rightarrow t = \frac{x}{20}$$

so

$$y = 6 + 30t - 5t^2$$

can be written as

$$y = 6 + 30 \times \frac{x}{20} - 5 \times \left(\frac{x}{20}\right)^2$$

$$y = 6 + 1.5x - \frac{x^2}{80}$$

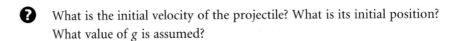 This is the cartesian equation

In this exercise make the simplification that g is 10 ms⁻².

1 Find the cartesian equation of the path of these projectiles by eliminating the parameter t.

(i) $x = 4t$ $y = 5t^2$

(ii) $x = 5t$ $y = 6 + 2t - 5t^2$

(iii) $x = 2 - t$ $y = 3t - 5t^2$

(iv) $x = 1 + 5t$ $y = 8 + 10t - 5t^2$

(v) $x = ut$ $y = 2ut - \frac{1}{2}gt^2$

2 A particle is projected with initial velocity 50 ms^{-1} at an angle of $36.9°$ to the horizontal. The point of projection is taken to be the origin, with the x axis horizontal and the y axis vertical in the plane of the particle's motion.

(i)   Show that at time t s, the height of the particle in metres is given by

$$y = 30t - 5t^2$$

and write down the corresponding expression for x.

(ii) Eliminate t between your equations for x and y to show that

$$y = \frac{3x}{4} - \frac{x^2}{320}.$$

(iii) Plot the graph of y against x using a scale of 2 cm for 10 m along both axes.

(iv) Mark on your graph the points corresponding to the position of the particle after 1, 2, 3, 4, ... seconds.

3 A golfer hits a ball with initial velocity 50 ms^{-1} at an angle α to the horizontal where $\sin \alpha = 0.6$.

(i)   Find the equation of its trajectory, assuming that air resistance may be neglected. The flight of the ball is recorded on film and its position vector, from the point where it was hit, is calculated. The unit vectors $\mathbf{i}$ and $\mathbf{j}$ are horizontal and vertical in the plane of the ball's motion. The results (to the nearest 0.5 m) are as follows:

Time (s)	Position (m)	Time (s)	Position (m)
0	$0\mathbf{i} + 0\mathbf{j}$	4	$152\mathbf{i} + 39\mathbf{j}$
1	$39.5\mathbf{i} + 24.5\mathbf{j}$	5	$187.5\mathbf{i} + 24.5\mathbf{j}$
2	$78\mathbf{i} + 39\mathbf{j}$	6	$222\mathbf{i} + 0\mathbf{j}$
3	$116.5\mathbf{i} + 44\mathbf{j}$		

(ii) On the same piece of graph paper draw the trajectory you found in part (i) and that found from analysing the film. Compare the two graphs and suggest a reason for any differences.

(iii) It is suggested that the horizontal component of the resistance to the motion of the golf ball is almost constant. Are the figures consistent with this?

Accessible points

The equation of the path of a projectile can be used to decide whether certain points can be reached by the projectile. The next two examples illustrate how this can be done.

EXAMPLE 6.5

A projectile is launched from the origin with an initial velocity $20\ \text{ms}^{-1}$ at an angle of $30°$ to the horizontal.

(i) Write down the position of the projectile after time t.

(ii) Show that the equation of the path is the parabola $y = 0.578x - 0.016x^2$.

(iii) Find y when $x = 3$.

(iv) Decide whether the projectile can hit a point 6 m above the ground.

SOLUTION

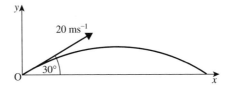

Figure 6.15

(i) Using horizontal and vertical components for the position:

Horizontally: $\qquad x = (20 \cos 30°)t$

Vertically: $\qquad y = (20 \sin 30°)t - 4.9t^2$

$\Rightarrow \qquad x = 17.3t$ ①

and $\qquad y = 10t - 4.9t^2$ ②

(ii) From equation ① $\qquad t = \dfrac{x}{17.3}$

Substituting this into equation ② for y gives

$$y = 10 \times \frac{x}{17.3} - 4.9 \times \frac{x^2}{17.3^2}$$

$\Rightarrow \qquad y = 0.578x - 0.016x^2$

(iii) When $x = 3$ $\qquad y = 0.578 \times 3 - 0.016 \times 9 = 1.59$

(iv) When $y = 6$ $\qquad 6 = 0.578x - 0.016x^2$

$\Rightarrow \qquad 0.016x^2 - 0.578x + 6 = 0$

In this quadratic equation the discriminant
$b^2 - 4ac = 0.578^2 - 4 \times 0.016 \times 6 = -0.0499$

You cannot find the square root of this negative number, so the equation cannot be solved for x. The projectile cannot hit a point 6 m above the ground.

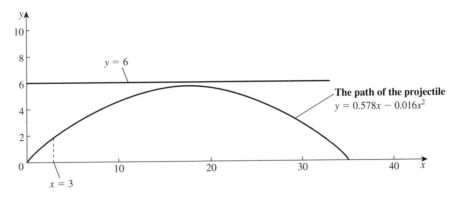

Figure 6.16

 How high does the projectile reach?

Looking at the sign of the discriminant of a quadratic equation is a good way of deciding whether points are within the range of a projectile. The example above is a quadratic in x. Example 6.7 involves a quadratic equation in $\tan \alpha$ but the same idea is used in part (iii) to determine whether certain points are within range.

EXAMPLE 6.7

A ball is hit from the origin with a speed of $14\,\text{ms}^{-1}$ at an angle α to the horizontal.

(i) Find the equation of the path of the ball in terms of $\tan \alpha$.

(ii) Find the two values of α which ensure that the ball passes through the point $(5, 2.5)$.

(iii) Decide whether the ball can pass through the points

    (a) $(10, 7.5)$ **(b)** $(8, 9)$.

SOLUTION

(i) The components of the initial velocity are $14 \cos \alpha$ and $14 \sin \alpha$, so the path of the ball is given by the equations

$$x = (14 \cos \alpha)t \text{ and } y = (14 \sin \alpha)t - 4.9t^2$$

$$\Rightarrow \qquad t = \frac{x}{14 \cos \alpha} \text{ and } y = 14 \sin \alpha \times \frac{x}{14 \cos \alpha} - 4.9 \times \frac{x^2}{(14 \cos \alpha)^2}$$

$$\Rightarrow \qquad y = x \tan \alpha - \frac{x^2}{40 \cos^2 \alpha}$$

You can then express this in a more useful form by using two trigonometrical identities

$$\frac{1}{\cos \alpha} = \sec \alpha \text{ and } \sec^2 \alpha = 1 + \tan^2 \alpha$$

So the equation of the path is $\qquad y = x \tan \alpha - \dfrac{1}{40} x^2 (1 + \tan^2 \alpha)$ ①

(ii) When $x = 5$ and $y = 2.5$ $\qquad 2.5 = 5 \tan \alpha - \dfrac{25}{40}(1 + \tan^2 \alpha)$

multiply by $\dfrac{40}{25}$ $\longrightarrow$ $\qquad 4 = 8 \tan \alpha - (1 + \tan^2 \alpha)$

$$\tan^2 \alpha - 8 \tan \alpha + 5 = 0$$

This is a quadratic in $\tan \alpha$ which has solution $\tan \alpha = \dfrac{8 \pm \sqrt{64 - 20}}{2}$.

So $\tan \alpha = 7.32$ or 0.68 giving possible angles of projection of $82°$ and $34°$.

(iii) (a) For the point $(10, 7.5)$, equation ① gives

$$7.5 = 10 \tan \alpha - \tfrac{10^2}{40}(1 + \tan^2 \alpha)$$

$$7.5 = 10 \tan \alpha - 2.5(1 + \tan^2 \alpha)$$
$$\tan^2 \alpha - 4 \tan \alpha + 4 = 0 \longleftarrow \qquad \boxed{\text{divide by 2.5}}$$

The discriminant $b^2 - 4ac = 16 - 16 = 0$

This zero discriminant means that the equation has two equal roots $\tan \alpha = 2$. The point $(10, 7.5)$ lies on the path of the projectile but there is only one possible angle of projection.

(b) For the point $(8, 9)$, equation ① gives $9 = 8 \tan \alpha - \tfrac{8^2}{40}(1 + \tan^2 \alpha)$

$$1.6 \tan^2 \alpha - 8 \tan \alpha + 10.6 = 0$$

The discriminant is $64 - 4 \times 1.6 \times 10.6 = -3.84$ and this has no real square root.

This equation cannot be solved to find $\tan \alpha$ so the projectile cannot pass through the point $(8, 9)$ if it has an initial speed of 14 ms^{-1}.

The diagram shows all the results.

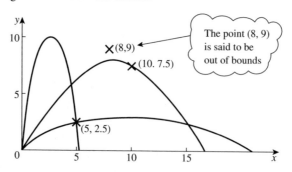

Figure 6.17

Use 10 ms^{-2} for g in this exercise unless otherwise instructed and use the modelling assumptions that air resistance can be ignored and the ground is horizontal.

1   A projectile is launched from the origin with an initial velocity 30 ms^{-1} at an angle of 45° to the horizontal.
   (i)   Write down the position of the projectile after time t.
   (ii) Show that the equation of the path is the parabola $y = x - 0.011x^2$.
   (iii) Find y when $x = 10$.
   (iv) Find x when $y = 20$.

2   Jack throws a cricket ball at a wicket 0.7 m high with velocity 10 ms^{-1} at 14° above the horizontal. The ball leaves his hand 1.5 m above the origin.
   (i)   Show that the equation of the path is the parabola
       $y = 1.5 + 0.25x - 0.053x^2$.
   (ii) How far from the wicket is he standing if the ball just hits the top?

3   While practising his tennis serve, Matthew hits the ball from a height of 2.5 m with a velocity of magnitude 25 ms^{-1} at an angle of 5° above the horizontal as shown in the diagram. In this question, take $g = 9.8$ ms^{-2}.

   (i)   Show that while in flight
       $y = 2.5 + 0.087x - 0.0079x^2$.
   (ii) Find the horizontal distance from the serving point to the spot where the ball lands.
   (iii) Determine whether the ball would clear the net, which is 1 m high and 12 m from the serving position in the horizontal direction.

4   Ching is playing volleyball. She hits the ball with initial speed u ms^{-1} from a height of 1 m at an angle of 35° to the horizontal.

   (i)   Define a suitable origin and x and y axes and find the equation of the trajectory of the ball in terms of x, y and u.

   The rules of the game require the ball to pass over the net, which is at height 2 m, and land inside the court on the other side, which is of length 5 m. Ching hits the ball straight along the court and is 3 m from the net when she does so.
   (ii) Find the minimum value of u for the ball to pass over the net.
   (iii) Find the maximum value of u for the ball to land inside the court.

5 A ball is hit from the origin with a speed of $10\,\text{ms}^{-1}$ at an angle α to the horizontal.

  (i) Show that the equation of the path is $y = x\tan\alpha - 0.05x^2\,(1 + \tan^2\alpha)$.

  (ii) Find the two values of α which ensure that the ball passes through the point $(2, 3)$.

  (iii) Decide whether the ball can pass through the points

    (a) $(5, 3)$ **(b)** $(3, 5)$.

6 A ball is hit from the point $(0, 2)$ with a speed of $40\,\text{ms}^{-1}$ at an angle α to the horizontal.

  (i) Find the equation of the path of the ball in terms of $\tan\alpha$.

  (ii) Find the two values of α which ensure that the ball passes through the point $(18, 0)$.

  (iii) Can the ball pass through the points   **(a)** $(160, 5)$ **(b)** $(150, 8)$.

7 Tennis balls are delivered from a machine at a height of 1 m and with a speed of $25\,\text{ms}^{-1}$. Dan stands 15 m from the machine and can reach to a height of 2.5 m.

  (i) Write down the equations for the horizontal and vertical displacements x and y after time t seconds for a ball which is delivered at an angle $\alpha°$ to the horizontal.

  (ii) Eliminate t from your equations to obtain the equation for y in terms of x and $\tan\alpha$.

  (iii) Can Dan reach a ball when $\tan\alpha$ is   **(a)** 0.5   **(b)** 0.2.

  (iv) Determine the values of α for which Dan can just hit the ball.

8 A high pressure hose is used to water a horizontal garden. The jet of water is modelled as a stream of small droplets (i.e. particles) projected at $15\,\text{ms}^{-1}$ at an angle α to the horizontal from a point 1.2 m above the garden. Air resistance may be neglected. The diagram shows this situation as well as the origin O and the axes. The unit of each axis is the metre.

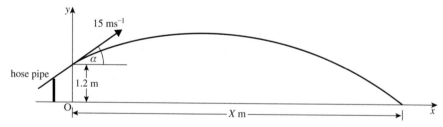

  (i) Show that the vertical height of a droplet at time t is given by

$$y = 1.2 + 15t\sin\alpha - 5t^2.$$

Write down the corresponding equation for the horizontal distance.

The horizontal range of a droplet projected at angle α is X m, as shown in the diagram.

  (ii) Use your answers from part (ii) to deduce that

$$X^2\tan^2\alpha - 45X\tan\alpha + X^2 - 54 = 0.$$

(iii) Show that it is possible to adjust the angle of projection so that the water lands on the point (20, 0) but that the point (30, 0) is out of range.

(iv) Does the maximum value of X occur when $\alpha = 45°$? Explain your answer briefly.

9 A golf ball is driven from the tee with speed $30\sqrt{2}$ ms^{-1} at an angle α to the horizontal.

(i) Show that during its flight the horizontal and vertical displacements x and y of the ball from the tee satisfy the equation

$$y = x \tan \alpha - \frac{x^2}{360}(1 + \tan^2 \alpha).$$

(ii) The golf ball just clears a tree 5 m high which is 150 m horizontally from the tee. Find the two possible values of $\tan \alpha$.

(iii) Use the discriminant of the quadratic equation in $\tan \alpha$ to find the greatest distance by which the golf ball can clear the tree and find the value of $\tan \alpha$ in this case.

(iv) The ball is aimed at the hole which is on the green immediately behind the tree. The hole is 160 m from the tee. What is the greatest height the tree could be without making it impossible to hit a hole in one? (Hint: $2 \sin \alpha \cos \alpha = \sin 2\alpha$)

10 A particle is projected up a slope of angle β where $\tan \beta = \frac{1}{2}$. The initial velocity of the particle is 50 ms^{-1} at angle α to the horizontal where $\sin \alpha = \frac{4}{5}$. The x and y axes are taken from the point of projection and are in the plane of the particle's motion as shown below.

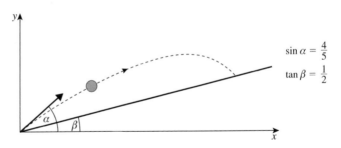

(i) Use values of g and $\tan \alpha$ to find the equation of the trajectory of the particle.

(ii) Explain why the equation of the slope is $y = \frac{1}{2}x$.

(iii) Solve the equations of the trajectory and the slope to find the co-ordinates of the point where the particle hits the slope.

(iv) What is the range of the particle up the slope?

Another particle is then projected from the same point and with the same initial speed but at an angle of $45°$ to the horizontal.

(v) Find the range of this particle up the slope.

(vi) Does an angle of projection of $45°$ to the horizontal result in the particle travelling the maximum distance up the slope?

General equations

The work done in this chapter can now be repeated for the general case using algebra. Assume a particle is projected from the origin with speed u at an angle α to the horizontal and that the only force acting on the particle is the force due to gravity. The x and y axes are horizontal and vertical through the origin, O, in the plane of motion of the particle.

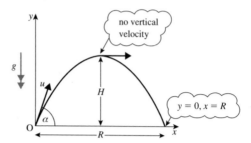

Figure 6.18

The components of velocity and position

	Horizontal motion		Vertical motion	
Initial position	0		0	
a	0		$-g$	
u	$u_x = u \cos \alpha$		$u_y = u \sin \alpha$	
v	$v_x = u \cos \alpha$	①	$v_y = u \sin \alpha - gt$	②
r	$x = ut \cos \alpha$	③	$y = ut \sin \alpha - \frac{1}{2} gt^2$	④

> $ut \cos \alpha$ is preferable to $u \cos \alpha t$ because this could mean $u \cos (\alpha t)$ which is incorrect

The maximum height

At its greatest height, the vertical component of velocity is zero.
From equation ②

$$u \sin \alpha - gt = 0$$

$$t = \frac{u \sin \alpha}{g}$$

Substitute in equation ④ to obtain the height of the projectile

$$y = u \times \frac{u \sin \alpha}{g} \times \sin \alpha - \frac{1}{2} g \times \frac{(u \sin \alpha)^2}{g^2}$$

$$= \frac{u^2 \sin^2 \alpha}{g} - \frac{u^2 \sin^2 \alpha}{2g}$$

The greatest height is $\qquad H = \dfrac{u^2 \sin^2 \alpha}{2g}$

The time of flight

When the projectile hits the ground, $y = 0$.

From equation ④:

$$y = ut \sin \alpha - \tfrac{1}{2}gt^2$$

$$0 = ut \sin \alpha - \tfrac{1}{2}gt^2$$

$$0 = t\left(u \sin \alpha - \tfrac{1}{2}gt\right)$$

The solution $t = 0$ is at the start of the motion

The time of flight is

$$t = \frac{2u \sin \alpha}{g}$$

The range

The range of the projectile is the value of x when $t = \dfrac{2u \sin \alpha}{g}$

From equation ③:

$$x = ut \cos \alpha$$

$$\Rightarrow \quad R = u \times \frac{2u \sin \alpha}{g} \times \cos \alpha$$

$$R = \frac{2u^2 \sin \alpha \cos \alpha}{g}$$

It can be shown that $2 \sin \alpha \cos \alpha = \sin 2\alpha$, so the range can be expressed as

$$R = \frac{u^2 \sin 2\alpha}{g}$$

The range is a maximum when $\sin 2\alpha = 1$, that is when $2\alpha = 90°$ or $\alpha = 45°$. The maximum possible horizontal range for projectiles with initial speed u is

$$R_{\text{max}} = \frac{u^2}{g}.$$

The equation of the path

From equation ③

$$t = \frac{x}{u \cos \alpha}$$

Substitute into equation ④ to give

$$y = u \times \frac{x}{u \cos \alpha} \times \sin \alpha - \tfrac{1}{2}g \times \frac{x^2}{(u \cos \alpha)^2}$$

$$y = x\frac{\sin \alpha}{\cos \alpha} - \frac{gx^2}{2u^2 \cos^2 \alpha}$$

$$y = x \tan \alpha - \frac{gx^2}{2u^2}(1 + \tan^2 \alpha)$$

 It is important that you understand the methods used to derive these formulae and don't rely on learning the results off by heart. They are only true when the given assumptions apply and the variables are as defined in figure 6.18.

 What are the assumptions on which this work is based?

The diagram shows how a wet ball projected on to a sloping table can be used to simulate a projectile.

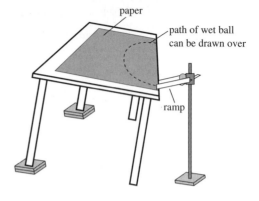

1 Ignoring rotation and friction, what is the ball's acceleration?
2 Does the mass of the ball affect the motion?

Set up the apparatus so that you can move the ramp to make different angles of projection with the same speed.

Figure 6.19

3 Can the same range be achieved using two different angles?
4 What angle gives the maximum range?
5 What is the shape of the curve containing all possible paths with the same initial speed?

FIREWORKS

A firework sends out sparks from ground level with the same speed, $20\,\text{ms}^{-1}$, in all directions. A spark starts at an angle α to the horizontal. Investigate the accessible points for this speed by plotting the trajectory for different values of α. (Use $10\,\text{ms}^{-2}$ for g).

Using a graphics calculator or other graph plotter investigate the shape of the curve which forms the outer limit for all possible sparks with trajectories which lie in a vertical plane.

Show that the trajectory of a spark is given by $y = x\tan\alpha - \dfrac{1}{80}x^2(1 + \tan^2\alpha)$.

A point (x, y) is on the boundary. Use the idea that the discriminant of the quadratic equation for $\tan\alpha$ is zero for the extreme points to find the equation of this curve (called the *bounding parabola*). Check your answer by plotting it on a diagram showing also several possible paths for the sparks.

1 Modelling assumptions for projectile motion with acceleration due to gravity

- a projectile is a particle
- it is not powered
- the air has no effect on its motion.

2 Projectile motion is usually considered in terms of horizontal and vertical components.

When the initial position is at O

Angle of projection $= \alpha$

Initial velocity, $\mathbf{u} = \begin{pmatrix} u \cos \alpha \\ u \sin \alpha \end{pmatrix}$

Acceleration, $\mathbf{g} = \begin{pmatrix} 0 \\ -g \end{pmatrix}$

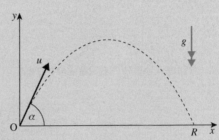

- At time t, velocity, $\mathbf{v} = \mathbf{u} + \mathbf{a}t$

$$\begin{pmatrix} v_x \\ v_y \end{pmatrix} = \begin{pmatrix} u \cos \alpha \\ u \sin \alpha \end{pmatrix} + \begin{pmatrix} 0 \\ -g \end{pmatrix} t$$

$$v_x = u \cos \alpha \qquad \textcircled{1}$$
$$v_y = u \sin \alpha - gt \qquad \textcircled{2}$$

- Displacement, $\mathbf{r} = \mathbf{u}t + \frac{1}{2}\mathbf{a}t^2$

$$\begin{pmatrix} x \\ y \end{pmatrix} = \begin{pmatrix} u \cos \alpha \\ u \sin \alpha \end{pmatrix} t + \frac{1}{2} \begin{pmatrix} 0 \\ -g \end{pmatrix} t^2$$

$$x = ut \cos \alpha \qquad \textcircled{3}$$
$$y = ut \sin \alpha - \frac{1}{2} gt^2 \qquad \textcircled{4}$$

3 At maximum height $v_y = 0$.
4 $y = 0$ when the projectile lands.
5 The time to hit the ground is twice the time to maximum height.

6 The equation of the path of a projectile is $y = x \tan \alpha - \dfrac{gx^2}{2u^2}(1 + \tan^2 \alpha)$.

7 For given u, x and y, this equation can be written as a quadratic in $\tan \alpha$.

When there are

- 2 roots the point is within bounds
- 1 root the point is at the limit
- 0 roots the point is out of bounds.

8 When the point of projection is (x_o, y_o) rather than $(0, 0)$

$$\mathbf{r} = \mathbf{r}_o + \mathbf{u}t + \frac{1}{2}\mathbf{a}t^2 \qquad \begin{pmatrix} x \\ y \end{pmatrix} = \begin{pmatrix} x_o \\ y_o \end{pmatrix} + \begin{pmatrix} u \cos \alpha \\ u \sin \alpha \end{pmatrix} t + \frac{1}{2} \begin{pmatrix} 0 \\ -g \end{pmatrix} t^2.$$

7 Forces and motion in two dimensions

Give me matter and motion and I will construct the Universe.

René Descartes

Finding resultant forces

❓ This cable car is stationary. Are the tensions in the cable greater than the weight of the car?

A child on a sledge is being pulled up a smooth slope of 20° by a rope which makes an angle of 40° with the slope. The mass of the child and sledge together is 20 kg and the tension in the rope is 170 N. Draw a diagram to show the forces acting on the child and sledge together. In what direction is the resultant of these forces?

When the child and sledge are modelled as a particle, all the forces can be assumed to be acting at a point. There is no friction force because the slope is smooth. Here is the force diagram.

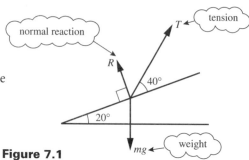

Figure 7.1

? The sledge is sliding along the slope. What direction is the resultant force acting on it?

You can find the normal reaction and the resultant force on the sledge using two methods.

Method 1: Using components

This method involves resolving forces into components in two perpendicular directions as in Chapter 5. It is easiest to use the components of the forces parallel and perpendicular to the slope in the directions of **i** and **j** as shown.

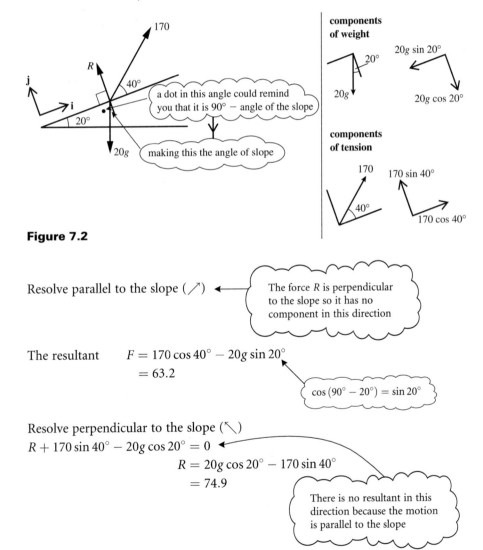

components of weight

components of tension

Figure 7.2

Resolve parallel to the slope ($\nearrow$) ← The force R is perpendicular to the slope so it has no component in this direction

The resultant $F = 170\cos 40° - 20g\sin 20°$
$= 63.2$

$\cos(90° - 20°) = \sin 20°$

Resolve perpendicular to the slope ($\nwarrow$)
$R + 170\sin 40° - 20g\cos 20° = 0$
$R = 20g\cos 20° - 170\sin 40°$
$= 74.9$

There is no resultant in this direction because the motion is parallel to the slope

The normal reaction is 75 N and the resultant 63 N up the slope.

Alternatively, you could have worked in column vectors as follows.

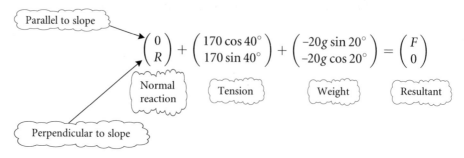

Parallel to slope

$$\begin{pmatrix} 0 \\ R \end{pmatrix} + \begin{pmatrix} 170 \cos 40° \\ 170 \sin 40° \end{pmatrix} + \begin{pmatrix} -20g \sin 20° \\ -20g \cos 20° \end{pmatrix} = \begin{pmatrix} F \\ 0 \end{pmatrix}$$

Normal reaction

Tension

Weight

Resultant

Perpendicular to slope

Note

Try resolving horizontally and vertically. You will obtain two equations in the two unknowns R and F. It is perfectly possible to solve these equations, but quite a lot of work. It is much easier to choose to resolve in directions which ensure that one component of at least one of the unknown forces is zero.

Once you know the resultant force, you can work out the acceleration of the sledge using Newton's second law.

$$F = ma$$
$$63.2 = 20a$$

The acceleration is $3.2 \, \text{ms}^{-2}$ (correct to 1 dp)

Method 2: Scale drawing

An alternative is to draw a scale diagram with the three forces represented by three of the sides of a quadrilateral taken in order (with the arrows following each other) as shown in figure 7.3. The resultant is represented by the fourth side AD. This must be parallel to the slope.

? In what order would you draw the lines in the diagram?

From the diagram you can estimate the normal reaction to be about 80 N and the resultant 60 N. This is a reasonable estimate, but components are more precise.

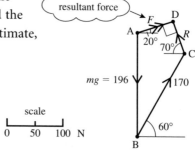

resultant force

$mg = 196$

scale

0   50   100 N

Figure 7.3

What can you say about the sledge in the cases when

(i) the length AD is not zero?

(ii) the length AD is zero so that the starting point on the quadrilateral is the same as the finishing point?

(iii) BC is so short that the point D is to the left of A as shown in the diagram?

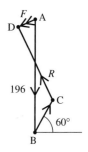

Figure 7.4

EXERCISE 7A

For each of the situations below, carry out the following steps. All forces are in newtons.

(i) Draw a scale diagram to show the polygon of the forces and the resultant.

(ii) State whether you think the forces are in equilibrium and, if not, estimate the magnitude and direction of the resultant.

(iii) Write the forces in component form, using the directions indicated and so obtain the components of the resultant.

Hence find the magnitude and direction of the resultant as in Chapter 5, page 94.

(iv) Compare your answers to parts (ii) and (iii).

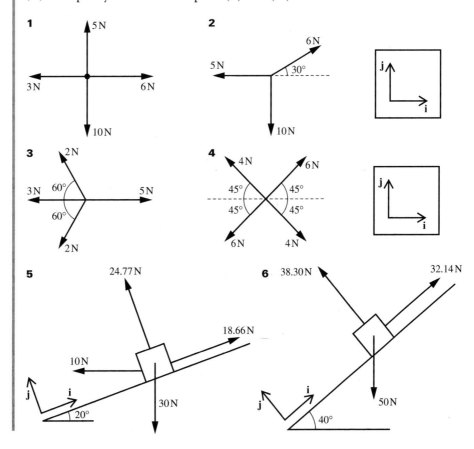

Forces in equilibrium

When forces are in equilibrium their vector sum is zero and the sum of their resolved parts in *any* direction is zero.

EXAMPLE 7.1

A brick of mass 3 kg is at rest on a rough plane inclined at an angle of 30° to the horizontal. Find the friction force F N, and the normal reaction R N of the plane on the brick.

SOLUTION

The diagram shows the forces acting on the brick.

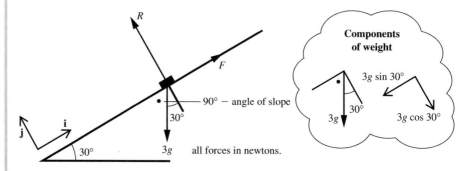

Figure 7.5

Take unit vectors **i** and **j** parallel and perpendicular to the plane as shown.
Since the brick is in equilibrium the resultant of the three forces acting on it is zero.

Resolving in the **i** direction $F - 29.4 \sin 30° = 0$ ⟵ $3g = 29.4$ ①

$$F = 14.7$$

Resolving in the **j** direction: $R - 29.4 \cos 30° = 0$ ②

$$R = 25.5$$

Written in vector form the equivalent is

$$\begin{pmatrix} F \\ 0 \end{pmatrix} + \begin{pmatrix} 0 \\ R \end{pmatrix} + \begin{pmatrix} -29.4 \sin 30° \\ -29.4 \cos 30° \end{pmatrix} = \begin{pmatrix} 0 \\ 0 \end{pmatrix}$$

Or alternatively

$$F\mathbf{i} + R\mathbf{j} - 29.4 \sin 30° \, \mathbf{i} - 29.4 \cos 30° \, \mathbf{j} = \mathbf{0}$$

Both these lead to the equations ① and ②.

The triangle of forces

When there are only three (non-parallel) forces acting and they are in equilibrium, the polygon of forces becomes a closed triangle as shown for the brick on the plane.

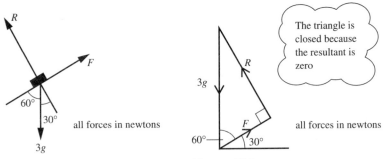

Figure 7.6   Figure 7.7

Then $\dfrac{F}{3g} = \cos 60°$

$$F = 29.4 \cos 60° = 14.7\,\text{N}$$

and similarly $R = 29.4 \sin 60° = 25.5\,\text{N}$

This is an example of the theorem known as the *triangle of forces*.

> When a body is in equilibrium under the action of three non-parallel forces, then
>
> **(i)** the forces can be represented in magnitude and direction by the sides of a triangle
> **(ii)** the lines of action of the forces pass through the same point.

When more than three forces are in equilibrium the first statement still holds but the triangle is then a polygon. The second is not necessarily true.

EXPERIMENT

The diagram shows two strings tied to a light ring at A and passing over smooth pulleys at B and C. Masses M_1 and M_2 are suspended from the free ends of the strings, and mass M is suspended from the ring at A. The whole system is mounted on a vertical board. Set up the apparatus as shown.

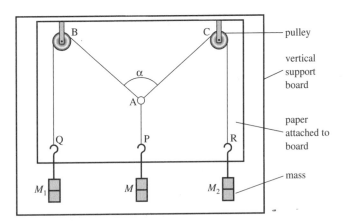

Figure 7.8

1 Before doing any experiments try to answer these questions intuitively.
   (i) Do the lengths of the strings affect the angles?
   (ii) For any given set of three masses, does the system have more than one equilibrium position?
   (iii) Is there a combination of masses for which
      (a) $\alpha = 90°$?   (b) $\alpha = 180°$?
   (iv) Is it always possible to find an equilibrium position?
   (v) When M_1 and M_2 are equal, what can you say about α when
      (a) $M = M_1$?   (b) $M > M_1$?   (c) $M < M_1$?

2 Now use the apparatus to investigate these questions. Model the system mathematically for masses M_1, M and M_2.
   (i) Make a list of the assumptions you could use in this situation.
   (ii) State which forces act at each point P, Q and R and on the ring at A.
   (iii) Write down the tensions T_1 in the left-hand string and T_2 in the right-hand string when the system is stationary.
   (iv) What will be the direction of the resultant of the tensions T_1 and T_2?

3 Check your model using three different masses.
   (i) Mark points A, B, C and P on the paper behind and join A to B, C and P.
   (ii) Measure the angles at A and mark the forces acting at A on your diagram.
   (iii) Draw the triangle of forces to find the resultant of T_1 and T_2.
   (iv) Resolve the forces T_1 and T_2 horizontally and vertically.

? Now look back at the questions in part 1 of the experiment. Are your results as you expected? If not, why not?

EXAMPLE 7.2

This example illustrates two methods for solving problems involving forces in equilibrium. With experience, you will find it easier to judge which method is best for a particular problem.

A sign of mass 10 kg is to be suspended by two strings arranged as shown in the diagram below. Find the tension in each string.

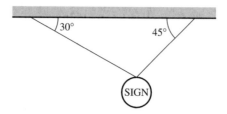

Figure 7.9

SOLUTION

The force diagram for this situation is given below.

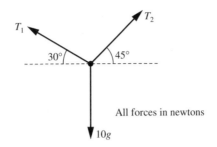

Figure 7.10

Method 1: Resolving forces

Vertically ($\uparrow$): $T_1 \sin 30° + T_2 \sin 45° - 10g = 0$

$$0.5T_1 + 0.707T_2 = 98 \qquad \text{①}$$

Horizontally ($\rightarrow$): $-T_1 \cos 30° + T_2 \cos 45° = 0$

$$-0.866T_1 + 0.707T_2 = 0 \qquad \text{②}$$

Subtracting ② from ① $1.366T_1 = 98$

$$T_1 = 71.74$$

Back substitution gives $T_2 = 87.87$

The tensions are 71.7 N and 87.9 N (to 1 dp).

Method 2: Triangle of forces

Since the three forces are in equilibrium they can be represented by the sides of a triangle taken in order.

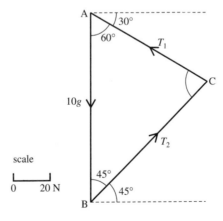

Figure 7.11

In what order would you draw the three lines in this diagram?

You can estimate the tensions by measurement. This will tell you that $T_1 \approx 72$ and $T_2 \approx 88$ in newtons.

Alternatively, you can use the sine rule to calculate T_1 and T_2 accurately.

In the triangle ABC, $\angle CAB = 60°$, $\angle ABC = 45°$ so $\angle BCA = 75°$ and so

$$\frac{T_1}{\sin 45°} = \frac{T_2}{\sin 60°} = \frac{98}{\sin 75°}$$

giving $$T_1 = \frac{98 \sin 45°}{\sin 75°}$$

and $$T_2 = \frac{98 \sin 60°}{\sin 75°}$$

As before the tensions are found to be $71.7\,\text{N}$ and $87.9\,\text{N}$.

? Lami's theorem states that when three forces acting at a point as shown in the diagram are in equilibrium then

$$\frac{F_1}{\sin \alpha} = \frac{F_2}{\sin \beta} = \frac{F_3}{\sin \gamma}.$$

Sketch a triangle of forces and say how the angles in the triangle are related to α, β and γ. Hence explain why Lami's theorem is true.

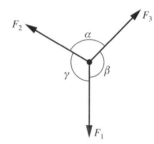

Figure 7.12

EXAMPLE 7.3

The picture shows three men involved in moving a packing case up to the top floor of a warehouse. Brian is pulling on a rope which passes round smooth pulleys at X and Y and is then secured to the point Z at the end of the loading beam.

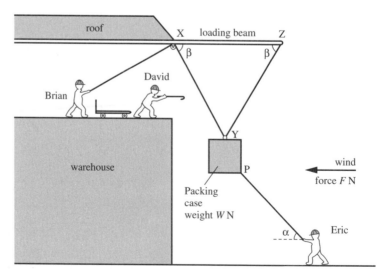

Figure 7.13

The wind is blowing directly towards the building. To counteract this, Eric is pulling on another rope, attached to the packing case at P, with just enough force and in the right direction to keep the packing case central between X and Z.

At the time of the picture the men are holding the packing case motionless.

(i) Draw a diagram showing all the forces acting on the packing case using T_1 and T_2 for the tensions in Brian and Eric's ropes, respectively.

(ii) Write down equations for the horizontal and vertical equilibrium of the packing case.

In one particular situation, W = 200, F = 50, $\alpha = 45°$ and $\beta = 75°$.

(iii) Find the tension T_1.

(iv) Explain why Brian has to pull harder if the wind blows stronger.

[MEI adapted]

SOLUTION

(i) The diagram shows all the forces acting on the packing case and the relevant angles.

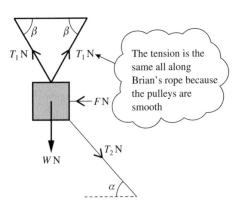

Figure 7.14 *Force diagram*

(ii) Equilibrium equations
Resolving horizontally ($\rightarrow$)

$$T_1 \cos \beta + T_2 \cos \alpha - F - T_1 \cos \beta = 0$$
$$T_2 \cos \alpha - F = 0 \qquad \text{①}$$

Resolving vertically ($\uparrow$)

$$T_1 \sin \beta + T_1 \sin \beta - T_2 \sin \alpha - W = 0$$
$$2T_1 \sin \beta - T_2 \sin \alpha - W = 0 \qquad \text{②}$$

(iii) When F = 50 and $\alpha = 45°$ equation ① gives

$$T_2 \cos 45° = 50$$
$$\Rightarrow T_2 \sin 45° = 50$$

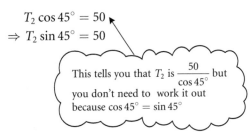

This tells you that T_2 is $\dfrac{50}{\cos 45°}$ but you don't need to work it out because $\cos 45° = \sin 45°$.

Substituting in ② gives $2T_1 \sin \beta - 50 - W = 0$

So when $W = 100$ and $\beta = 75°$ $2T_1 \sin 75° = 150$

$$T_1 = \frac{150}{2 \sin 75°}$$

The tension in Brian's rope is $77.65\,\text{N} = 78\,\text{N}$ (to the nearest N).

(iv) When the wind blows harder, F increases. Given that all the angles remain unchanged, Eric will have to pull harder so the vertical component of T_2 will increase. This means that T_1 must increase and Brian must pull harder.

> Or $F = T_2 \cos \alpha$, so as F increases, T_2 increases $\Rightarrow T_2 \sin \alpha + W$ increases $\Rightarrow 2T_1 \sin \beta$ increases. Hence T_1 increases.

EXERCISE 7B

1 The picture shows a boy, Halley, holding onto a post while his two older sisters, Sheuli and Veronica, try to pull him away.

Taking **i** and **j** to be unit vectors in perpendicular horizontal directions the forces, in newtons, exerted by the two girls are:

Sheuli	$24\mathbf{i} + 18\mathbf{j}$
Veronica	$25\mathbf{i} + 60\mathbf{j}$

(i) Calculate the magnitude and direction of the force of each of the girls.

(ii) Use a scale drawing to estimate the magnitude and direction of the resultant of the forces exerted by the two girls.

(iii) Write the resultant in terms of **i** and **j** and so calculate (to 3 significant figures) its magnitude and direction.

Check that your answers agree with those obtained by scale drawing in part (ii).

2 The diagram shows a girder CD of mass 20 tonnes being held stationary by a crane (which is not shown). The rope from the crane (AB) is attached to a ring at B. Two ropes, BC and BD, of equal length attach the girder to B; the tensions in each of these ropes is $T\,\text{N}$.

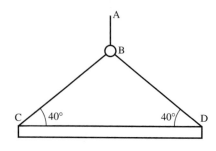

(i) Draw a diagram showing the forces acting on the girder.

(ii) Write down, in terms of T, the horizontal and vertical components of the tensions in the ropes acting at C and D.

(iii) Hence show that the tension in the rope BC is 152.5 kN (to 1 dp).

(iv) Draw a diagram to show the three forces acting on the ring at B.

(v) Hence calculate the tension in the rope AB.

(vi) How could you have known the answer to part (v) without any calculations?

3 The diagram shows a simple model of a crane. The structure is at rest in a vertical plane. The rod and cables are of negligible mass and the load suspended from the joint at A is 30 N.

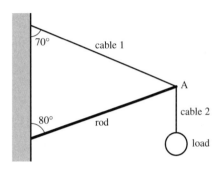

(i) Draw a diagram showing the forces acting on

    (a) the load   **(b)** the joint at A.

(ii) Calculate the forces in the rod and cable 1 and state whether they are in compression or in tension.

4 An angler catches a very large fish. When he tries to weigh it he finds that it is more than the 10 kg limit of his spring balance. He borrows another spring balance of exactly the same design and uses the two to weigh the fish, as shown in diagram A. Both balances read 8 kg.

(i) What is the mass of the fish?

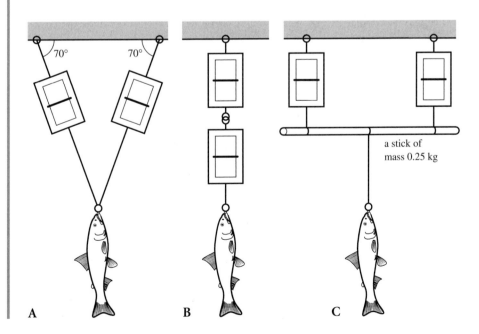

The angler believes the mass of the fish is a record and asks a witness to confirm it. The witness agrees with the measurements but cannot follow the calculations. He asks the angler to weigh the fish in two different positions, still using both balances. These are shown in diagrams B and C.

Assuming the spring balances themselves to have negligible mass, state the readings of the balances as set up in

(ii) diagram B

(iii) diagram C.

(iv) Which of the three methods do you think is the best?

5 The diagram shows a device for crushing scrap cars. The light rod AB is hinged at A and raised by a cable which runs from B round a pulley at D and down to a winch at E. The vertical strut EAD is rigid and strong and AD = AB. A weight of mass 1 tonne is suspended from B by the cable BC. When the weight is correctly situated above the car it is released and falls on to the car.

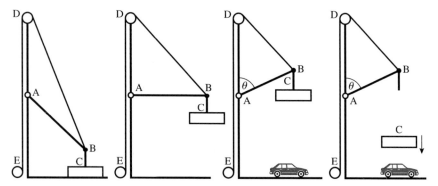

Just before the weight is released the rod AB makes angle θ with the upward vertical AD and the weight is at rest.

(i) Draw a diagram showing the forces acting at point B in this position.

(ii) Explain why the rod AB must be in thrust and not in tension.

(iii) Draw a diagram showing the vector sum of the forces at B (i.e. the polygon of forces).

(iv) Calculate each of the three forces acting at B when

    (a) $\theta = 90°$ **(b)** $\theta = 60°$.

6 Four wires, all of them horizontal, are attached to the top of a telegraph pole as shown in the plan view on the right. The pole is in equilibrium and tensions in the wires are as shown.

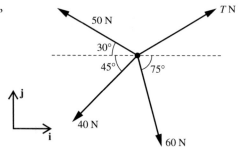

(i) Using the unit vectors **i** and **j** shown in the diagram, show that the force of 60 N may be written as $15.5\mathbf{i} - 58.0\mathbf{j}$ N (to 3 significant figures).

(ii) Find T in
   (a) component form   (b) magnitude and direction form.
(iii) The force T is changed to $40\mathbf{i} + 35\mathbf{j}$ N. Show that there is now a resultant force on the pole and find its magnitude and direction.

7 A ship is being towed by two tugs. Each tug exerts forces on the ship as indicated. There is also a drag force on the ship.

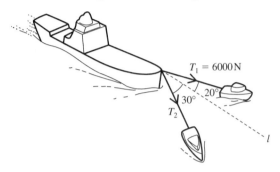

(i)   Write down the components of the tensions in the towing cables along and perpendicular to the line of motion, l, of the ship.
(ii) There is no resultant force perpendicular to the line l. Find T_2.
(iii) The ship is travelling with constant velocity along the line l. Find the magnitude of the drag force acting on it.

8 A boat of mass 500 kg is being winched up a beach which slopes at $10°$ to the horizontal. The maximum friction between the boat and the beach is 3500 N and the rope from the boat to the winch is parallel to the slope of the beach. The boat is on the point of moving up the beach.
(i)   Draw a diagram showing all the forces acting on the boat.
(ii) Write all these forces in components parallel and perpendicular to the slope.
(iii) Find the tension in the rope.

A little later the boat is moving at a constant speed of $1\,\mathrm{cm\,s}^{-1}$.
(iv) What is the tension in the rope now?

The rope breaks.
(v) Will the boat ever start to slide back down the slope?

9 A skier of mass 50 kg is skiing straight down a $15°$ slope.
(i)   Draw a diagram showing the forces acting on the skier.
(ii) Resolve these forces into components parallel and perpendicular to the slope.

The skier is travelling at constant speed.
(iii) Find the normal reaction of the slope on the skier and the resistance force on her.

The skier later returns to the top of the slope by being pulled up it at constant speed by a rope parallel to the slope.
(iv) Assuming the resistance on the skier is the same as before, calculate the tension in the rope.

10 The diagram shows a block of
mass 5 kg on a rough inclined
plane. The block is attached to
a 3 kg weight by a light string
which passes over a smooth
pulley and is on the point of
sliding up the slope.

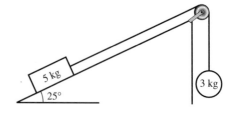

(i) Draw a diagram showing the forces acting on the block.
(ii) Resolve these forces into components parallel and perpendicular to the
slope.
(iii) Find the force of resistance to the block's motion.

The 3 kg weight is replaced by one of mass m kg.
(iv) Find the value of m for which the block is on the point of sliding down
the slope, assuming the resistance to motion is the same as before.

11 Two husky dogs are pulling a
sledge. They both exert forces of
60 N but at different angles to the
line of the sledge, as shown in the
diagram. The sledge is moving
straight forwards.

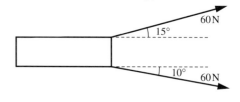

(i) Resolve the two forces into components parallel and perpendicular to
the line of the sledge.
(ii) Hence find
(a) the overall forward force from the dogs
(b) the sideways force.

The resistance to motion is 20 N along the line of the sledge but up to 400 N
perpendicular to it.
(iii) Find the magnitude and direction of the overall horizontal force on the
sledge.
(iv) How much force is lost due to the dogs not pulling straight
forwards?

12 The diagram shows a man suspended by
means of a rope which is attached at one
end to a peg at a fixed point A on a vertical
wall and at the other to a belt round his
waist. The man has weight $80g$ N, the
tension in the rope is T and the reaction
of the wall on the man is R. The rope is
inclined at 35° to the vertical and R is
inclined at $\alpha°$ to the vertical as shown.
The man is in equilibrium.

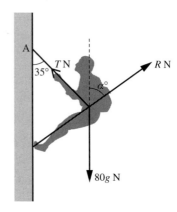

(i) Explain why $R > 0$.

(ii) By considering his horizontal and vertical equilibrium separately, obtain two equations connecting T, R and α.

(iii) Given that $\alpha = 45$, show that T is about 563 N and find R.

(iv) What is the magnitude and the direction of the force on the peg at A?

The peg at A is replaced by a smooth pulley. The rope is passed over the pulley and tied to a hook at B directly below A. Calculate

(v) the new value of the tension in the rope section BA

(vi) the magnitude of the force on the pulley at A.

[MEI]

13

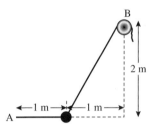

One end of a string of length 1 m is fixed to a mass of 1 kg and the other end is fixed to a point A. Another string is fixed to the mass and passes over a frictionless pulley at B which is 1 m horizontally from A but 2 m above it. The tension in the second string is such that the mass is held at the same horizontal level as the point A.

(i) Show that the tension in the horizontal string fixed to the mass and to A is 4.9 N and find the tension in the string which passes over the pulley at B. Find also the angle that this second string makes with the horizontal.

(ii) If the tension in this second string is slowly increased by drawing more of it over the pulley at B describe the path followed by the mass. Will the point A, the mass, and the point B, ever lie in a straight line? Give reasons for your answer.

14

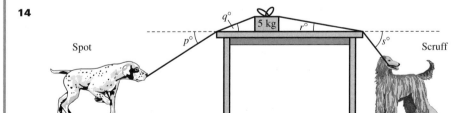

A parcel of mass 5 kg is on a horizontal table and has two ribbons attached as shown. The table is smooth and offers no frictional resistance to the parcel or the ribbons. Spot and Scruff are two dogs pulling on the ribbons. $p > q$ and $s > r$ at all times so that the ribbons remain in contact with the edges of the table. Initially, the parcel is in equilibrium in the middle of the table ($q = r$) and Spot pulls with a force of 60 N.

(i) Why must Scruff also pull with a force of 60 N?

(ii) Draw a force diagram for the parcel and calculate the reaction of the table on the parcel if $q = r = 40$.

The parcel is moved and reaches a new position of equilibrium with $q = 30$ and $r = 45$. The ribbons remain in contact with the edge of the table and Spot continues to pull with a force of 60 N.

(iii) With what force does Scruff now pull?

(iv) With Spot still pulling with a force of 60 N, express the force with which Scruff pulls in terms of q and r. Explain why the closer the parcel is to Scruff the harder he has to pull to maintain equilibrium.

[MEI]

15

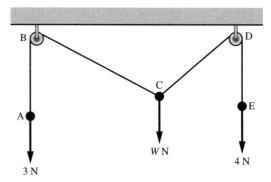

3 N

The diagram shows three small objects of weights 3, W and 4 newtons tied to the points A, C and E respectively on a light inextensible string. The string passes over small smooth pulleys at points B and D, which are on the same horizontal line. The weight W is such that, in equilibrium, C always lies between B and D. Initially the system is in equilibrium with $W = 5$. (The diagram should not be taken to represent this particular case.)

(i) State the tensions in the string sections AB, BC, DE and DC.

(ii) Draw a diagram to show how the forces acting on the object at C may be represented by the sides of a triangle and hence, or otherwise, calculate the angles which BC and CD make with the vertical.

(iii) Draw a sketch diagram showing the pulleys at B and D and the position of C.

Mark in the values of the angles CBD and CDB and the forces acting on the object at C. The system can be in equilibrium for other values of W. By means of further diagrams of the types produced in parts (ii) and (iii),

(iv) explain why W cannot be as great as 7,

(v) explain why W cannot be as small as $\sqrt{7}$.

[Hint: $3^2 + 7 = 4^2$]

[MEI]

Newton's second law in two dimensions

When the forces acting on an object are not in equilibrium it will have an acceleration and you can use Newton's second law to solve problems about its motion.

The equation $\mathbf{F} = m\mathbf{a}$ is a vector equation. The resultant force acting on a particle is equal in both magnitude and direction to the mass × acceleration. It can be written in components as

$$\begin{pmatrix} F_1 \\ F_2 \end{pmatrix} = m \begin{pmatrix} a_1 \\ a_2 \end{pmatrix} \text{ or } F_1\,\mathbf{i} + F_2\,\mathbf{j} = m\,(a_1\,\mathbf{i} + a_2\,\mathbf{j})$$

so that $F_1 = ma_1$ and $F_2 = ma_2$.

❓ What direction is the resultant force acting on a child sliding on a sledge down a smooth straight slope inclined at $15°$ to the horizontal?

EXAMPLE 7.4

Sam and his sister are sledging, but Sam wants to ride by himself. His sister gives him a push at the top of a smooth straight $15°$ slope and lets go when he is moving at $2\,\text{ms}^{-1}$. He continues to slide for 5 seconds before using his feet to produce a braking force of 95 N parallel to the slope. This brings him to rest. Sam and his sledge have a mass of 30 kg.

How far does he travel altogether?

SOLUTION

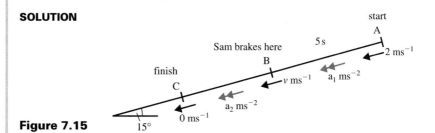

Figure 7.15

To answer this question, you need to know Sam's acceleration for the two parts of his journey. These are constant so you can then use the constant acceleration formulae.

Sliding freely

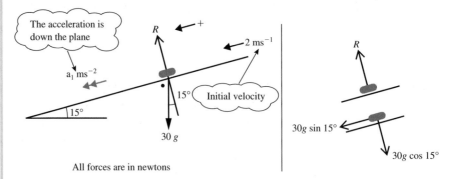

All forces are in newtons

Figure 7.16

Using Newton's second law in the direction of the acceleration gives

$$30g \sin 15° = 30a_1$$
$$a_1 = 2.54$$

Resultant force down the plane = mass × acceleration

Now you know a_1 you can find how far Sam slides (s_1 m) and his speed ($v \, \mathrm{ms^{-1}}$) before braking.

$$s = ut + \tfrac{1}{2}at^2$$
$$s_1 = 2 \times 5 + 1.27 \times 25 = 41.75$$
$$v = u + at$$
$$v = 2 + 2.54 \times 5 = 14.7$$

Given $u = 2$, $t = 5$, $a = 2.54$

Braking

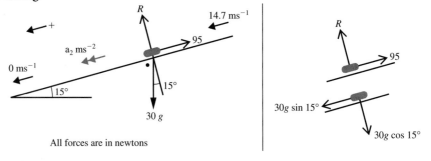

All forces are in newtons

Figure 7.17

By Newton's second law down the plane

$$\text{Resultant force} = \text{mass} \times \text{acceleration}$$
$$30g \sin 15° - 95 = 30a_2$$
$$a_2 = -0.63$$
$$v^2 = u^2 + 2as$$
$$0 = 14.7^2 - 2 \times 0.63 \times s_2$$

Given $u = 14.7$, $v = 0$, $a = -0.63$

$$s_2 = \frac{14.7^2}{1.26} = 171.5$$

Sam travels a total distance of $41.75 + 171.5 \, \mathrm{m} = 213 \, \mathrm{m}$ to the nearest metre.

? Make a list of the modelling assumptions used in this example. What would be the effect of changing these?

EXAMPLE 7.5

A skier is being pulled up a smooth 25° dry ski slope by a rope which makes an angle of 35° with the horizontal. The mass of the skier is 75 kg and the tension in the rope is 350 N. Initially the skier is at rest at the bottom of the slope. The slope is smooth. Find the skier's speed after 5 s and find the distance he has travelled in that time.

SOLUTION

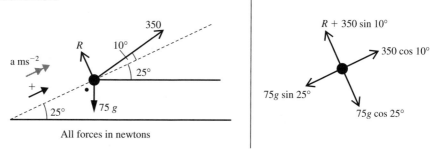

All forces in newtons

Figure 7.18

In the diagram the skier is modelled as a particle. Since the skier moves parallel to the slope consider motion in that direction.

$$\text{Resultant force} = \text{mass} \times \text{acceleration}$$

$$350 \cos 10° - 75g \sin 25° = 75 \times a$$

Taking g as 9.8 $\longrightarrow$ $a = \dfrac{34.06}{75} = 0.454 \text{ (to 3 dp)}.$

This is a constant acceleration so use the constant acceleration formulae.

$$v = u + at$$
$$v = 0 + 0.454 \times 5 \quad\longleftarrow\quad u = 0,\ a = 0.454,\ t = 5$$
$$\text{Speed} = 2.27 \text{ ms}^{-1} \text{ (to 2 dp)}.$$

$$s = ut + \tfrac{1}{2} at^2$$
$$s = 0 + \tfrac{1}{2} \times 0.454 \times 25$$
$$\text{Distance travelled} = 5.68 \text{ m (to 2 dp)}.$$

EXAMPLE 7.6

A car of mass 1000 kg, including its driver, is being pushed along a horizontal road by three people as indicated in the diagram. The car is moving in the direction PQ.

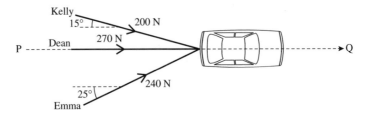

Figure 7.19

(i) Calculate the total force exerted by the three people in the direction PQ.

(ii) Calculate the force exerted overall by the three people in the direction perpendicular to PQ.

(iii) Explain briefly why the car does not move in the direction perpendicular to PQ.

Initially the car is stationary and 5 s later it has a speed of $2\,\text{ms}^{-1}$ in the direction PQ.

(iv) Calculate the force of resistance to the car's movement in the direction PQ assuming the three people continue to push as described above.

[MEI, part]

SOLUTION

(i) Resolving in the direction PQ, the components in newtons are:

Kelly $200\cos 15° = 193$
Dean   270
Emma $240\cos 25° = 218$

Total force in the direction PQ = 681 N.

(ii) Resolving perpendicular to PQ ($\uparrow$) the components are:

Kelly $-200\sin 15° = -51.8$
Dean   0
Emma $240\sin 25° = 101.4$

Total force in the direction perpendicular to PQ = 49.6 N.

(iii) The car does not move perpendicular to PQ because the force in this direction is balanced by a sideways (lateral) friction force between the tyres and the road.

(iv) To find the acceleration, $a\,\text{ms}^{-2}$, of the car:

$$v = u + at$$
$$2 = 0 + 5a \quad \longleftarrow \quad u = 0, v = 2, t = 5$$
$$a = 0.4$$

When the resistance to motion in the direction QP is $R\,\text{N}$, the diagram shows all the horizontal forces acting on the car and its acceleration.

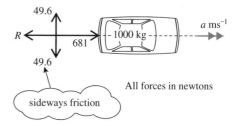

The weight of the car is in the third dimension, perpendicular to this plane and is balanced by the normal reaction of the ground

sideways friction

All forces in newtons

Figure 7.20

The resultant force in the direction PQ is $(681 - R)$ N. So by Newton II

$$681 - R = 1000a$$
$$R = 681 - 400$$

The resistance to motion in the direction PQ is 281 N.

1 The forces $\mathbf{F}_1 = 4\mathbf{i} - 5\mathbf{j}$ and $\mathbf{F}_2 = 2\mathbf{i} + \mathbf{j}$, in newtons, act on a particle of mass 4 kg.

   (i) Find the acceleration of the particle in component form.

   (ii) Find the magnitude of the particle's acceleration.

2 Two forces $\mathbf{P}_1$ and $\mathbf{P}_2$ act on a particle of mass 2 kg giving it an acceleration of $5\mathbf{i} + 5\mathbf{j}$ (in ms^{-2}).

   (i) If $\mathbf{P}_1 = 6\mathbf{i} - \mathbf{j}$ (in newtons), find $\mathbf{P}_2$.

   (ii) If instead $\mathbf{P}_1$ and $\mathbf{P}_2$ both act in the same direction but $\mathbf{P}_1$ is four times as big as $\mathbf{P}_2$ find both forces.

3 The diagram shows a girl pulling a sledge at steady speed across level snow-covered ground using a rope which makes an angle of 30° to the horizontal. The mass of the sledge is 8 kg and there is a resistance force of 10 N.

   (i) Draw a diagram showing the forces acting on the sledge.

   (ii) Find the magnitude of the tension in the rope.

The girl comes to an area of ice where the resistance force on the sledge is only 2 N. She continues to pull the sledge with the same force as before and with the rope still taut at 30°.

   (iii) What acceleration must the girl have in order to do this?

   (iv) How long will it take to double her initial speed of 0.4 ms^{-1}?

4 The picture shows a situation which has arisen between two anglers, Davies and Jones, standing at the ends of adjacent jetties. Their lines have become entangled under the water with the result that they have both hooked the same fish, which has mass 1.9 kg. Both are reeling in their lines as hard as they can in order to claim the fish.

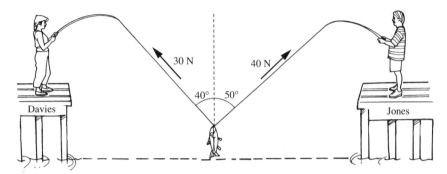

(i) Draw a diagram showing the forces acting on the fish.

(ii) Resolve the tensions in both anglers' lines into horizontal and vertical components and so find the total force acting on the fish.

(iii) Find the magnitude and direction of the acceleration of the fish.

(iv) At this point Davies' line breaks. What happens to the fish?

5 A crate of mass 30 kg is being pulled up a smooth slope inclined at $30°$ to the horizontal by a rope which is parallel to the slope. The crate has acceleration $0.75 \, \text{ms}^{-2}$.

(i) Draw a diagram showing the forces acting on the crate and the direction of its acceleration.

(ii) Resolve the forces in directions parallel and perpendicular to the slope.

(iii) Find the tension in the rope.

(iv) The rope suddenly snaps. What happens to the crate?

6 A cyclist of mass 60 kg rides a cycle of mass 7 kg. The greatest forward force that she can produce is 200 N but she is subject to air resistance and friction totalling 50 N.

(i) Draw a diagram showing the forces acting on the cyclist when she is going uphill.

(ii) What is the angle of the steepest slope that she can ascend?

The cyclist reaches a slope of $8°$ with a speed of $5 \, \text{ms}^{-1}$ and rides as hard as she can up it.

(iii) Find her acceleration and the distance she travels in 5 s.

(iv) What is her speed now?

7 A builder is demolishing the chimney of a house and slides the old bricks down to the ground on a straight chute 10 m long inclined at $42°$ to the horizontal. Each brick has mass 3 kg.

(i) Draw a diagram showing the forces acting on a brick as it slides down the chute, assuming the chute to have a flat cross section and a smooth surface.

(ii) Find the acceleration of the brick.

(iii) Find the time the brick takes to reach the ground.

In fact the chute is not smooth and the brick takes 3 s to reach the ground.

(iv) Find the frictional force acting on the brick, assuming it to be constant.

8 Two students are lowering a trunk down a rough slope inclined at 25° to the horizontal using a single rope which is parallel to the slope, as shown in the diagram. Using W for the weight of the trunk, T for the tension in the rope, F for the frictional force between the trunk and the ground and R for the normal reaction of the ground on the trunk.

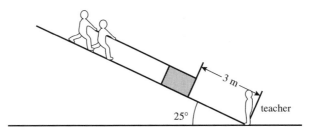

(i) draw a diagram showing all the external forces acting on the trunk.

The two students have a combined mass of 110 kg and can just prevent the trunk from sliding down the slope by pulling with a force which is, in magnitude, 20% of their combined **weight**.

(ii) Show that the tension in the rope is about 216 N.

The trunk has a mass of 100 kg.

(iii) Show that the frictional force acting on the trunk is about 199 N.

The students lose their grip and let go of the rope. The trunk slides down the slope toward a teacher 3 m away.

(iv) Assuming the frictional force is 199 N whilst the trunk is sliding, calculate the acceleration of the trunk and the time the teacher has to get out of the way after the trunk starts to slide.

[MEI]

9 A box of mass 80 kg is to be pulled along a horizontal floor by means of a light rope. The rope is pulled with a force of 100 N and the rope is inclined at 20° to the horizontal, as shown in the diagram.

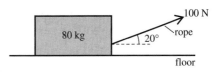

(i) Explain briefly why the box cannot be in equilibrium if the floor is smooth.

In fact the floor is not smooth and the box is in equilibrium.

(ii) Draw a diagram showing all the external forces acting on the box.

(iii) Calculate the frictional force between the box and the floor and also the normal reaction of the floor on the box, giving your answers correct to three significant figures.

The maximum value of the frictional force between the box and the floor is 120 N and the box is now pulled along the floor with the rope always inclined at 20° to the horizontal.

(iv) Calculate the force with which the rope must be pulled for the box to move at a constant speed. Give your answer correct to three significant figures.

(v) Calculate the acceleration of the box if the rope is pulled with a force of 140 N.

[MEI]

10 Railway trucks in coal mines were sometimes pulled by men. One such truck is standing on a straight, horizontal section of track. The man pulls on a light inextensible rope AB. The rope is horizontal and at an angle θ to the direction of the track. The man walks parallel to the track.

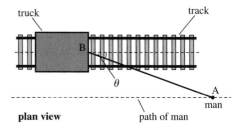

Initially the magnitude of the tension in the rope is 100 N and $\theta = 10°$. This tension in the rope is not enough to move the truck from rest.

(iii) Calculate the components of the tension in the rope parallel and perpendicular to the track.

(ii) What force prevents the truck from moving perpendicular to the track?

(iii) What is the magnitude of the resistance to the forward motion of the truck

The man now pulls harder to move the truck. The truck moves from rest against a resistance to its forward motion of $(100 + 44 \sin \theta)$ N at all times. There is a constant tension in the rope and θ has the constant value of 10°. It takes the man 15 s to reach his normal walking speed of 1.5 ms^{-1}.

(iv) Explain briefly why the acceleration of the truck is constant. With what force must the man pull on the rope to maintain the speed of 1.5 ms^{-1}? How far does the man walk before he reaches his normal walking speed?

(v) In order to avoid an obstacle, the man follows a path in which θ is increased. Assuming that the force in the rope does not change, what effect does this have on the motion of the truck?

[MEI]

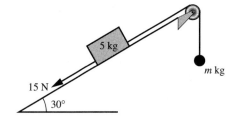

A block of mass 5 kg is at rest on a plane which is inclined at 30° to the horizontal. A light, inelastic string is attached to the block, passes over a smooth pulley and supports a mass m which is hanging freely. The part of the string between the block and the pulley is parallel to a line of greatest slope of the plane. A friction force of 15 N opposes the motion of the block. The diagram shows the block when it is slipping up the plane at a constant speed.

Give your answers correct to two significant figures.

(i)   Copy the diagram and mark in all the forces acting on the block and the hanging mass, including the tension in the string.

(ii) Calculate the value of m when the block slides up the plane at a constant speed and find the tension in the string.

(iii) Calculate the acceleration of the system when $m = 6$ kg and find the tension in the string in this case.

[MEI]

12 A tile of mass 2 kg slides 1.2 m from rest down a roof inclined at 30° to the horizontal against a resistance of 5 N. The edge of the roof is 3 m above the ground.

(i)   Draw a diagram to show the forces acting on the tile as it slides down the roof and use it to find its acceleration.

Calculate also

(ii) the velocity of the tile when it reaches the end of the roof.

(iii) the time taken for the tile to hit the ground.

(iv) the horizontal distance from the end of the roof to the point where the tile lands on the ground.

13 Charlotte is sliding down a water chute at the swimming pool and enters the last part at a speed of 4 ms^{-1}. The chute is smooth and straight and the last part has length 2.5 m. It is inclined at 40° to the horizontal and its end is 1.5 m above the water.

(i)   Find Charlotte's acceleration down the chute. Does it depend on her mass?

(ii) What is her speed as she comes to the end of the chute?

(iii) What horizontal distance from the end of the chute does Charlotte hit the water?

7

1 The forces acting on a particle can be combined to form a *resultant force* using scale drawing (trigonometry) or components.

Scale drawing

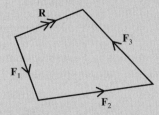

- Draw an accurate diagram, then measure the resultant. This is less accurate than calculation
- To calculate the resultant, find the resultant of two forces at a time using the trigonometry of triangles. This is time-consuming for more than two forces.

Components

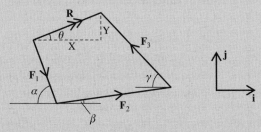

When $\mathbf{R}$ is $X\mathbf{i} + Y\mathbf{j}$

$$X = F_1 \cos \alpha + F_2 \cos \beta - F_3 \cos \gamma$$

$$Y = -F_1 \sin \alpha + F_2 \sin \beta + F_3 \sin \gamma$$

$$|\mathbf{R}| = \sqrt{X^2 + Y^2}$$

$$\tan \theta = \frac{Y}{X}$$

2 **Equilibrium**

When the resultant $\mathbf{R}$ is zero, the forces are in equilibrium.

3 **Triangle of forces**

If a body is in equilibrium under three non-parallel forces, their lines of action are concurrent and they can be represented by a triangle.

4 **Newton's second law**

When the resultant $\mathbf{R}$ is not zero there is an acceleration $\mathbf{a}$ and $\mathbf{R} = m\mathbf{a}$

5 When a particle is on a slope, it is usually helpful to resolve in directions parallel and perpendicular to the slope.

8 General motion

The goal of applied mathematics is to understand reality mathematically.

G. G. Hall

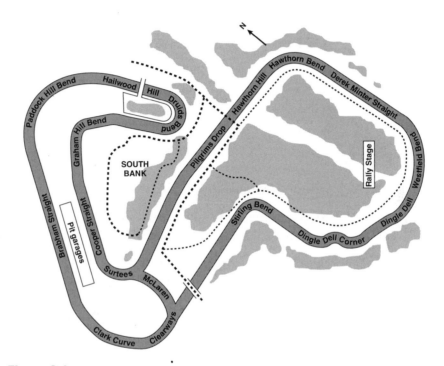

Figure 8.1

So far you have studied motion with constant acceleration in a straight line, but the motion of a car round the Brand's Hatch racing circuit shown in the diagram is much more complex. In this chapter you will see how to deal first with variable acceleration and later with motion in two dimensions.

Motion in one dimension

The equations you have used for constant acceleration do not apply when the acceleration varies. You need to go back to first principles.

Consider how displacement, velocity and acceleration are related to each other. The velocity of an object is the rate at which its position changes with time. When the velocity is not constant the position–time graph is a curve.

The rate of change of the position is the gradient of the tangent to the curve. You can find this by differentiating.

$$v = \frac{ds}{dt} \qquad ①$$

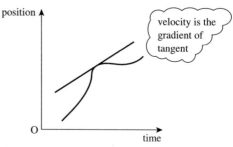

velocity is the gradient of tangent

Figure 8.2

Similarly, the acceleration is the rate at which the velocity changes, so

$$a = \frac{dv}{dt} = \frac{d^2s}{dt^2} \qquad ②$$

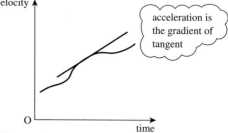

acceleration is the gradient of tangent

Figure 8.3

Using differentiation

When you are given the position of a moving object in terms of time, you can use equations ① and ② to solve problems even when the acceleration is not constant.

EXAMPLE 8.1

An object moves along a straight line so that its position at time t in seconds is given by

$$x = 2t^3 - 6t \text{ (in metres)} \quad (t \geqslant 0).$$

(i) Find expressions for the velocity and acceleration of the object at time t.
(ii) Find the values of x, v and a when $t = 0$, 1, 2 and 3.
(iii) Sketch the graphs of x, v and a against time.
(iv) Describe the motion of the object.

SOLUTION

(i)   Position $x = 2t^3 - 6t$ ①

Velocity $v = \dfrac{dx}{dt} = 6t^2 - 6$ ②

Acceleration $a = \dfrac{dv}{dt} = 12t$ ③

You can now use these three equations to solve problems about the motion of the object.

(ii) When

$t =$	0	1	2	3
From ①	$x =$	0	−4	4	36
From ②	$v =$	−6	0	18	48
From ③	$a =$	0	12	24	36

(iii) The graphs are drawn under each other so that you can see how they relate.

(iv) The object starts at the origin and moves towards the negative direction, gradually slowing down.

At $t = 1$ it stops instantaneously and changes direction, returning to its initial position at about $t = 1.7$.

It then continues moving in the positive direction with increasing speed.

The acceleration is increasing at a constant rate. This cannot go on for much longer or the speed will become excessive.

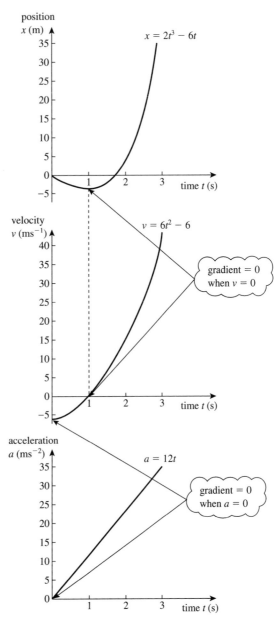

Figure 8.4

1 In each of the following cases

(i) $s = 10 + 2t - t^2$

(ii) $s = -4t + t^2$

(iii) $x = t^3 - 5t^2 + 4$

    (a) find expressions for the velocity;

    (b) use your equations to write down the initial position and velocity;

    (c) find the time and position when the velocity is zero.

2 In each of the following cases
   (i) $v = 4t + 3$
   (ii) $v = 6t^2 - 2t + 1$
   (iii) $v = 7t - 5$
      (a) find expressions for the acceleration;
      (b) use your equations to write down the initial velocity and acceleration.

3 The distance travelled by a cyclist is modelled by

$$s = 4t + 0.5t^2 \text{ in S.I. units.}$$

Find expressions for the velocity and the acceleration of the cyclist at time t.

4 In each of the following cases
   (i) $x = 15t - 5t^2$
   (ii) $x = 6t^3 - 18t^2 - 6t + 3$
      (a) find expressions for the velocity and the acceleration;
      (b) draw the acceleration–time graph and, below it, the velocity–time graph with the same scale for time and the origins in line;
      (c) describe how the two graphs for each object relate to each other;
      (d) describe how the velocity and acceleration change during the motion of each object.

Finding displacement from velocity

How can you find an expression for the position of an object when you know its velocity in terms of time?

One way of thinking about this is to remember that $v = \dfrac{ds}{dt}$, so you need to do the opposite of differentiation, that is integrate, to find s.

$$s = \int v \, dt$$

The dt indicates that you must write v in terms of t before integrating

EXAMPLE 8.2

The velocity (in ms^{-1}) of a model train which is moving along straight rails is

$$v = 0.3t^2 - 0.5$$

Find its displacement from its initial position

(i) after time t
(ii) after 3 seconds.

SOLUTION

(i) The displacement at any time is $s = \int v \, dt$
$$= \int (0.3t^2 - 0.5) \, dt$$
$$= 0.1t^3 - 0.5t + c$$

To find the train's displacement from its initial position, put $s = 0$ when $t = 0$. This gives $c = 0$ and so $s = 0.1t^3 - 0.5t$.

You can use this equation to find the displacement at any time before the motion changes.

(ii) After 3 seconds, $t = 3$ and $s = 2.7 - 1.5$.

The train is 1.2 m from its initial position.

 When using integration don't forget the constant. This is very important in mechanics problems and you are usually given some extra information to help you find the value of the constant.

The area under a velocity–time graph

In Chapter 1 you saw that the area under a velocity–time graph represents a displacement. Both the area under the graph and the displacement are found by integrating. To find a particular displacement you calculate the area under the velocity–time graph by integration using suitable limits.

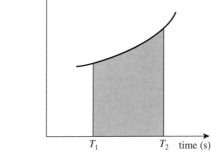

Figure 8.5

The distance travelled between the times T_1 and T_2 is shown by the shaded area on the graph.

$$s = \text{area} = \int_{T_1}^{T_2} v \, dt$$

EXAMPLE 8.3

A car moves between two sets of traffic lights, stopping at both. Its speed $v \, \text{ms}^{-1}$ at time t s is modelled by

$$v = \frac{1}{20} t (40 - t), \quad 0 \leqslant t \leqslant 40.$$

Find the times at which the car is stationary and the distance between the two sets of traffic lights.

SOLUTION

The car is stationary when $v = 0$. Substituting this into the expression for the speed gives

$$0 = \frac{1}{20} t (40 - t)$$

$$t = 0 \text{ or } t = 40.$$

These are the times when the car starts to move away from the first set of traffic lights and stops at the second set.

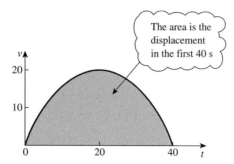

Figure 8.6

The distance between the two sets of lights is given by

$$\text{Distance} = \int_0^{40} \tfrac{1}{20} t \,(40 - t)\, dt$$

$$= \tfrac{1}{20} \int_0^{40} 40t - t^2 \, dt$$

$$= \tfrac{1}{20} \left[20t^2 - \tfrac{t^3}{3} \right]_0^{40}$$

$$= 533.\dot{3} \, m$$

Finding velocity from acceleration

You can also find the velocity from the acceleration by using integration.

$$a = \tfrac{dv}{dt}$$

$$\Rightarrow \quad v = \int a \, dt$$

The next example shows how you can obtain equations for motion using integration.

EXAMPLE 8.4

The acceleration of a particle (in ms^{-2}) at time t seconds is given by

$$a = 6 - t.$$

The particle is initially at the origin with velocity -2 ms^{-1}. Find an expression for

(i) the velocity of the particle after t s
(ii) the position of the particle after t s.

Hence find the velocity and position 6 s later.

SOLUTION

The information given may be summarised as follows:

at $t = 0$, $s = 0$ and $v = -2$;

at time t, $a = 6 - t$.　①

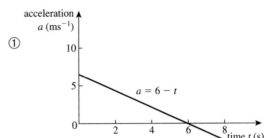

(i) $\dfrac{dv}{dt} = a = 6 - t$

Integrating gives

$$v = 6t - \tfrac{1}{2}t^2 + c$$

When $t = 0$, $v = -2$

so　$-2 = 0 - 0 + c$

　　$c = -2$

at time t

$$v = 6t - \tfrac{1}{2}t^2 - 2 \qquad ②$$

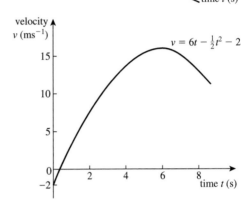

(ii) $\dfrac{ds}{dt} = v = 6t - \dfrac{1}{2}t^2 - 2$

Integrating gives

$$s = 3t^2 - \tfrac{1}{6}t^3 - 2t + k$$

When $t = 0$, $s = 0$

so　$0 = 0 - 0 - 0 + k$

　　$k = 0$

At time t

$$s = 3t^2 - \tfrac{1}{6}t^3 - 2t \qquad ③$$

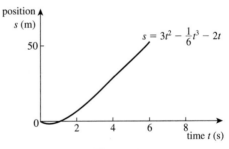

Figure 8.7

 Notice that two different arbitrary constants (c and k) are necessary when you integrate twice. You could call them c_1 and c_2 if you wish.

The three numbered equations can now be used to give more information about the motion in a similar way to the *suvat* equations. (The *suvat* equations only apply when the acceleration is constant.)

When $t = 6$　　$v = 36 - 18 - 2 = 16$　　　　　　　　　　from ②

When $t = 6$　　$s = 108 - 36 - 12 = 60$　　　　　　　　from ③

The particle has a velocity of $+16\,\text{ms}^{-1}$ and is at $+60\,\text{m}$ after 6 s.

1 Find expressions for the position in each of these cases.

   (i) $v = 4t + 3$; initial position 0.

   (ii) $v = 6t^3 - 2t^2 + 1$; when $t = 0$, $s = 1$.

   (iii) $v = 7t^2 - 5$; when $t = 0$, $s = 2$.

2 The speed of a ball rolling down a hill is modelled by $v = 1.7t$ (in ms^{-1}).

   (i) Draw the speed–time graph of the ball.

   (ii) How far does the ball travel in 10 s?

3 Until it stops moving, the speed of a bullet t s after entering water is modelled by $v = 216 - t^3$ (in ms^{-1}).

   (i) When does the bullet stop moving?

   (ii) How far has it travelled by this time?

4 During braking the speed of a car is modelled by $v = 40 - 2t^2$ (in ms^{-1}) until it stops moving.

   (i) How long does the car take to stop?

   (ii) How far does it move before it stops?

5 In each case below, the object moves along a straight line with acceleration a in ms^{-2}. Find an expression for the velocity v (ms^{-1}) and position x (m) of each object at time t s.

   (i) $a = 10 + 3t - t^2$; the object is initially at the origin and at rest.

   (ii) $a = 4t - 2t^2$; at $t = 0$, $x = 1$ and $v = 2$.

   (iii) $a = 10 - 6t$; at $t = 1$, $x = 0$ and $v = -5$.

The constant acceleration equations revisited

? In which of the cases in question 1 above is the acceleration constant? Which constant acceleration equations give the same results for s, v and a in this case? Why would the constant acceleration equations not apply in the other two cases?

You can use integration to prove the equations for constant acceleration. When a is constant (and only then)

$$v = \int a \, dt = at + c_1$$

When $t = 0$, $v = u$ $\qquad\qquad\qquad u = 0 + c_1$

$\Rightarrow \qquad v = u + at$ $\qquad\qquad\qquad\qquad$ ①

You can integrate this again to find $s = ut + \frac{1}{2}at^2 + c_2$

> u and a are both constant

If $s = s_0$ when $t = 0$, $c_2 = s_0$ and $\qquad s = ut + \frac{1}{2}at^2 + s_0$ $\qquad$ ②

How can you use these to derive the other equations for constant acceleration?

$$s = \tfrac{1}{2}(u+v)t + s_0 \qquad \text{③}$$

$$v^2 - u^2 = 2a(s - s_0) \qquad \text{④}$$

$$s = vt - \tfrac{1}{2}at^2 + s_0 \qquad \text{⑤}$$

1 A boy throws a ball up in the air from a height of 1.5 m and catches it at the same height. Its height in metres at time t seconds is

$$y = 1.5 + 15t - 5t^2.$$

  (i) What is the vertical velocity $v\,\text{ms}^{-1}$ of the ball at time t?

  (ii) Find the position, velocity and speed of the ball at $t = 1$ and $t = 2$.

  (iii) Sketch the position–time, velocity–time and speed–time graphs for $0 \leqslant t \leqslant 3$.

  (iv) When does the boy catch the ball?

  (v) Explain why the distance travelled by the ball is not equal to $\int_0^3 v\,\mathrm{d}t$ and state what information this expression does give.

2 An object moves along a straight line so that its position in metres at time t seconds is given by

$$x = t^3 - 3t^2 - t + 3 \quad (t \geqslant 0).$$

  (i) Find the position, velocity and speed of the object at $t = 2$.

  (ii) Find the smallest time when

  (a) the position is zero

  (b) the velocity is zero.

  (iii) Sketch position–time, velocity–time and speed–time graphs for $0 \leqslant t \leqslant 3$.

  (iv) Describe the motion of the object.

3 Two objects move along the same straight line. The velocities of the objects (in ms^{-1}) are given by $v_1 = 16t - 6t^2$ and $v_2 = 2t - 10$ $(t \geqslant 0)$.

Initially the objects are 32 m apart. At what time do they collide?

4 An object moves along a straight line so that its acceleration in metres per second is given by $a = 4 - 2t$. It starts its motion at the origin with speed $4\,\text{ms}^{-1}$ in the direction of x increasing.

  (i) Find as functions of t the velocity and position of the object.

  (ii) Sketch the position–time, velocity–time and acceleration–time graphs for $0 \leqslant t \leqslant 2$.

  (iii) Describe the motion of the object.

5 Nick watches a golfer putting her ball 24 m from the edge of the green and into the hole and he decides to model the motion of the ball. Assuming that the ball is a particle travelling along a straight line he models its distance, s metres, from the golfer at time t seconds by

$$s = -\frac{3}{2}t^2 + 12t \quad 0 \leqslant t \leqslant 4.$$

(i) Find the value of s when $t = 0$, 1, 2, 3 and 4.
(ii) Explain the restriction $0 \leqslant t \leqslant 4$.
(iii) Find the velocity of the ball at time t seconds.
(iv) With what speed does the ball enter the hole?
(v) Find the acceleration of the ball at time t seconds.

6 Andrew and Elizabeth are having a race over 100 m. Their accelerations (in ms^{-2}) are as follows:

Andrew		Elizabeth	
$a = 4 - 0.8t$	$0 \leqslant t \leqslant 5$	$a = 4$	$0 \leqslant t \leqslant 2.4$
$a = 0$	$t > 5$	$a = 0$	$t > 2.4$

(i) Find the greatest speed of each runner.
(ii) Sketch the speed–time graph for each runner.
(iii) Find the distance Elizabeth runs while reaching her greatest speed.
(iv) How long does Elizabeth take to complete the race?
(v) Who wins the race, by what time margin and by what distance?

On another day they race over 120 m, both running in exactly the same manner.
(vi) What is the result now?

7 Christine is a parachutist. On one of her descents her vertical speed, $v\,ms^{-1}$, t s after leaving an aircraft is modelled by

$$\begin{aligned} v &= 8.5t & 0 \leqslant t \leqslant 10 \\ v &= 5 + 0.8\,(t - 20)^2 & 10 < t \leqslant 20 \\ v &= 5 & 20 < t \leqslant 90 \\ v &= 0 & t > 90 \end{aligned}$$

(i) Sketch the speed–time graph for Christine's descent and explain the shape of each section.
(ii) How high is the aircraft when Christine jumps out?
(iii) Write down expressions for the acceleration during the various phases of Christine's descent. What is the greatest magnitude of her acceleration?

8 A train starts from rest at a station. Its acceleration is shown on the acceleration–time graph below.

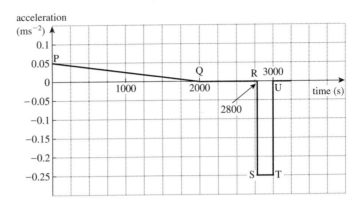

(i) Describe what is happening during the phases of the train's journey represented by the lines PQ, QR and ST.

(ii) The equation of the line PQ is of the form $a = mt + c$. Find the values of the constants m and c.

(iii) Find the maximum speed of the train.

(iv) What is the speed of the train when $t = 3000$?

(v) How far does the train travel during the first 3000 s?

9 A man of mass 70 kg is standing in a lift which, at a particular time, has an acceleration of $1.6\,\text{ms}^{-2}$ upwards. He is holding a parcel of mass 5 kg by a single string.

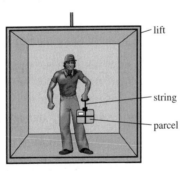

(i) Draw a diagram marking the forces acting on the parcel and the direction of the acceleration.

(ii) Show that the tension in the string is 57 N.

(iii) Calculate the reaction of the lift floor on the man.

During the first two seconds after starting from rest, the lift has acceleration in ms^{-2} modelled by $3t\,(2 - t)$, where t is in seconds. The maximum tension the string can withstand is 60 N.

(iv) By investigating the maximum acceleration of the system, or otherwise, determine whether the string will break during this time.

[MEI]

10

A————————0——————————————————B

An object moves along a straight line AB so that its displacement s metres from 0 at time t seconds is given by

$$s = t^3 - 3t \quad \text{for } 0 \leqslant t \leqslant 2$$

The displacement is positive in the direction OB.

(i) Find the velocity and acceleration of the object at time t.

(ii) When is the velocity zero?

(iii) Sketch the velocity–time and acceleration–time graphs for the motion for $0 \leqslant t \leqslant 2$.

(iv) Describe the motion of the object for $0 \leqslant t \leqslant 2$.

(v) Calculate the total distance travelled by the object from $t = 0$ to $t = 2$.

[MEI]

11 A bird leaves its nest for a short horizontal flight along a straight line and then returns. Michelle models its distance, s metres, from the nest at time t seconds by

$$s = 25t - \tfrac{5}{2}t^2, \quad 0 \leqslant t \leqslant 10.$$

(i) Find the value of s when $t = 2$.

(ii) Explain the restriction $0 \leqslant t \leqslant 10$.

(iii) Find the velocity of the bird at time t seconds.

(iv) What is the greatest distance of the bird from the nest?

(v) Michelle's teacher tells her that a better model would be

$$s = 10t^2 - 2t^3 + \tfrac{1}{10}t^4$$

Show that the two models agree about the time of the journey and the greatest distance travelled. Compare their predictions about velocity and suggest why the teacher's model is better.

[MEI]

12 A battery-operated toy dog starts at a point O and moves in a straight line. Its motion is modelled by the velocity–time graph below.

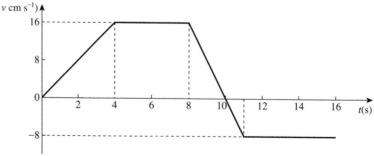

(i) Calculate the displacement from O of the toy

(a) after 10 seconds   **(b)** after 16 seconds.

(ii) Write down expressions for the velocity of the toy at time t seconds in the intervals $0 \leqslant t \leqslant 4$ and $4 \leqslant t \leqslant 8$.

(iii) Obtain expressions for the displacement from O of the toy at time t seconds in the intervals $0 \leqslant t \leqslant 4$ and $4 \leqslant t \leqslant 8$.

An alternative model for the motion of the toy in the interval $0 \leqslant t \leqslant 10$ is $v = \frac{2}{3}(10t - t^2)$, where v is the velocity in $cm\,s^{-1}$.

(iv) Calculate the difference in the displacement from O after 10 seconds as predicted by the two models.

[MEI]

Motion in two and three dimensions

In your work on projectile motion you have met the idea that the position of an object can be represented by a vector

$$\mathbf{r} = \begin{pmatrix} x \\ y \end{pmatrix} \quad \text{or} \quad x\mathbf{i} + y\mathbf{j}.$$

When a ball is thrown into the air, for example, its position might be given by

$$\mathbf{r} = \begin{pmatrix} 5t \\ 12t - 5t^2 \end{pmatrix}$$

so in this case $x = 5t$ and $y = 12t - 5t^2$. You can plot the path of the ball by finding the cartesian equation as in Chapter 6, page 118 or by finding the values of $\mathbf{r}$ and hence x and y for several values of t.

t	0	0.5	1	1.5	2	2.4
$\mathbf{r}$	$\begin{pmatrix} 0 \\ 0 \end{pmatrix}$	$\begin{pmatrix} 2.5 \\ 4.75 \end{pmatrix}$	$\begin{pmatrix} 5 \\ 7 \end{pmatrix}$	$\begin{pmatrix} 7.5 \\ 6.75 \end{pmatrix}$	$\begin{pmatrix} 10 \\ 4 \end{pmatrix}$	$\begin{pmatrix} 12 \\ 0 \end{pmatrix}$

? Why is 2.4 chosen as the last value for t?

Figure 8.8 shows the path of the ball and also its position $\mathbf{r}$ when $t = 2$.

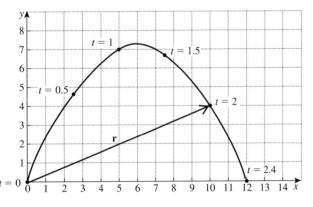

Figure 8.8

Finding the velocity and acceleration

The equation for **r** can be differentiated to give the velocity and acceleration.

When
$$\mathbf{r} = \begin{pmatrix} 5t \\ 12t - 5t^2 \end{pmatrix}$$

$$\mathbf{v} = \begin{pmatrix} 5 \\ 12 - 10t \end{pmatrix}$$

$$\mathbf{a} = \begin{pmatrix} 0 \\ -10 \end{pmatrix}$$

Note

The direction of the acceleration is not at all obvious when you look at the diagram of the path of the ball.

Newton's notation

When you write derivatives in column vectors, the notation becomes very cumbersome so many people use Newton's notation when differentiating with respect to time. In this notation a dot is placed over the variable for each differentiation. For example $\dot{x} = \frac{dx}{dt}$ and $\ddot{x} = \frac{d^2x}{dt^2}$.

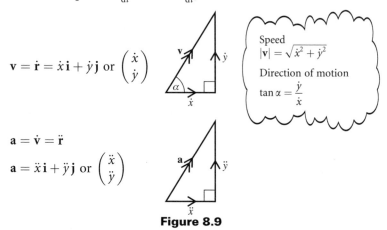

$$\mathbf{v} = \dot{\mathbf{r}} = \dot{x}\mathbf{i} + \dot{y}\mathbf{j} \text{ or } \begin{pmatrix} \dot{x} \\ \dot{y} \end{pmatrix}$$

Speed
$$|\mathbf{v}| = \sqrt{\dot{x}^2 + \dot{y}^2}$$

Direction of motion
$$\tan \alpha = \frac{\dot{y}}{\dot{x}}$$

$$\mathbf{a} = \dot{\mathbf{v}} = \ddot{\mathbf{r}}$$

$$\mathbf{a} = \ddot{x}\mathbf{i} + \ddot{y}\mathbf{j} \text{ or } \begin{pmatrix} \ddot{x} \\ \ddot{y} \end{pmatrix}$$

Figure 8.9

The next example shows how you can use these results.

EXAMPLE 8.5

Relative to an origin on a long, straight beach, the position of a speedboat is modelled by the vector

$$\mathbf{r} = (2t + 2)\mathbf{i} + (12 - t^2)\mathbf{j}$$

where **i** and **j** are unit vectors perpendicular and parallel to the beach. Distances are in metres and the time t is in seconds.

(i) Calculate the distance of the boat from the origin, O, when the boat is 6 m from the beach.

(ii) Sketch the path of the speedboat for $0 \leqslant t \leqslant 3$.

(iii) Find expressions for the velocity and acceleration of the speedboat at time t. Is the boat ever at rest? Explain your answer.

(iv) For $t = 3$, calculate the speed of the boat and the angle its direction of motion makes to the line of the beach.

(v) Suggest why this model for the motion of the speedboat is unrealistic for large t.

[MEI]

SOLUTION

(i) $\mathbf{r} = (2t + 2)\,\mathbf{i} + (12 - t^2)\,\mathbf{j}$ so the boat is 6 m from the beach when
$$x = 2t + 2 = 6$$
Then $t = 2$ and $y = 12 - t^2 = 8$

The distance from O is $\sqrt{6^2 + 8^2} = 10\,\text{m}$.

(ii) The table shows the position at different times and the path of the boat is shown on the graph in figure 8.10.

t	0	1	2	3
$\mathbf{r}$	$2\mathbf{i} + 12\mathbf{j}$	$4\mathbf{i} + 11\mathbf{j}$	$6\mathbf{i} + 8\mathbf{j}$	$8\mathbf{i} + 3\mathbf{j}$

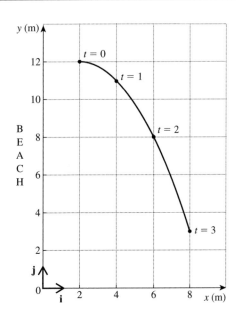

Figure 8.10

(iii) $$\mathbf{r} = (2t + 2)\,\mathbf{i} + (12 - t^2)\,\mathbf{j}$$
$$\Rightarrow \quad \mathbf{v} = \dot{\mathbf{r}} = 2\mathbf{i} - 2t\,\mathbf{j}$$
$$\text{and} \quad \mathbf{a} = \dot{\mathbf{v}} = -2\mathbf{j}$$

The boat is at rest if both components of velocity ($\dot{x}$ and $\dot{y}$) are zero at the same time. But $\dot{x}$ is always 2, so the velocity can never be zero.

(iv) When $t = 3$ $\mathbf{v} = 2\mathbf{i} - 6\mathbf{j}$

The angle $\mathbf{v}$ makes with the beach is α as shown where

$$\tan \alpha = \frac{2}{6}$$

$$\alpha = 18.4°$$

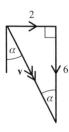

(v) According to this model, the speed after time t is

$$|\mathbf{v}| = |2\mathbf{i} - 2t\,\mathbf{j}| = \sqrt{2^2 + (-2t)^2} = \sqrt{4 + 4t^2}$$ **Figure 8.11**

As t increases, the speed increases at an increasing rate so there must come a time when the boat is incapable of going at the predicted speed and the model cannot then apply.

 Notice that the direction of motion is found using the velocity and not the position.

Using integration

When you are given the velocity or acceleration and wish to work backwards to the displacement, you need to integrate. The next two examples show how you can do this with vectors.

EXAMPLE 8.6

An aircraft is dropping a crate of supplies on to level ground. Relative to an observer on the ground, the crate is released at the point with position vector $\begin{pmatrix} 650 \\ 576 \end{pmatrix}$ m and with initial velocity $\begin{pmatrix} -100 \\ 0 \end{pmatrix}$ ms^{-1}, where the directions are horizontal and vertical. Its acceleration is modelled by

$$\mathbf{a} = \begin{pmatrix} -t + 12 \\ \frac{1}{2}t - 10 \end{pmatrix} \text{ for } t \leqslant 12 \text{ s.}$$

(i) Find an expression for the velocity vector of the crate at time t.
(ii) Find an expression for the position vector of the crate at time t.
(iii) Verify that the crate hits the ground 12 s after its release and find how far from the observer this happens.

SOLUTION

(i)
$$\mathbf{a} = \frac{d\mathbf{v}}{dt} = \begin{pmatrix} -t + 12 \\ \frac{1}{2}t - 10 \end{pmatrix} \quad ①$$

Integrating gives $\mathbf{v} = \begin{pmatrix} -\frac{1}{2}t^2 + 12t + c_1 \\ \frac{1}{4}t^2 - 10t + c_2 \end{pmatrix}$

> You can treat horizontal and vertical motion separately if you wish

At $t = 0$ $\mathbf{v} = \begin{pmatrix} -100 \\ 0 \end{pmatrix}$ $\Rightarrow$ $\begin{array}{l} 0 + 0 + c_1 = -100 \\ 0 - 0 + c_2 = 0 \end{array}$

Velocity $\mathbf{v} = \begin{pmatrix} -\frac{1}{2}t^2 + 12t - 100 \\ \frac{1}{4}t^2 - 10t \end{pmatrix}$ ②

(ii)
$$\mathbf{v} = \frac{d\mathbf{r}}{dt}$$

Integrating again gives
$$\mathbf{r} = \begin{pmatrix} -\frac{1}{6}t^3 + 6t^2 - 100t + k_1 \\ \frac{1}{12}t^3 - 5t^2 + k_2 \end{pmatrix}$$

At $t = 0$
$$\mathbf{r} = \begin{pmatrix} 650 \\ 576 \end{pmatrix} \quad \Rightarrow \quad \begin{matrix} k_1 = 650 \\ k_2 = 576 \end{matrix}$$

Position vector
$$\mathbf{r} = \begin{pmatrix} -\frac{1}{6}t^3 + 6t^2 - 100t + 650 \\ \frac{1}{12}t^3 - 5t^2 + 576 \end{pmatrix}$$ ③

(iii) At $t = 12$
$$\mathbf{r} = \begin{pmatrix} -\frac{1}{6} \times 12^3 + 6 \times 12^2 - 100 \times 12 + 650 \\ \frac{1}{12} \times 12^3 - 5 \times 12^2 + 576 \end{pmatrix}$$

$$\mathbf{r} = \begin{pmatrix} 26 \\ 0 \end{pmatrix}$$

Since $y = 0$, the crate hits the ground after 12 s and it is then $x = 26$ m in front of the observer.

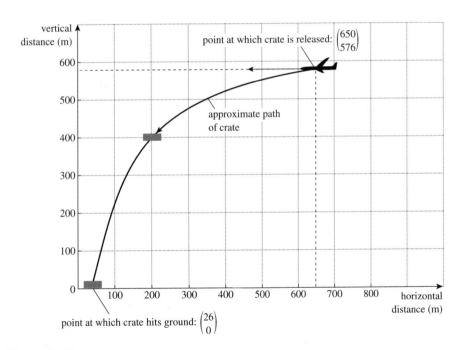

Figure 8.12

Note

When you integrate a vector in two dimensions you need a constant of integration for each direction, for example c_1 and c_2 as above.

It is also a good idea to number your equations for **a**, **v** and **r** so that you can find them easily if you want to use them later.

Force as a function of time

When the force acting on an object is given as a function of t you can use Newton's second law to find out about its motion. You can now write this as $\mathbf{F} = m\mathbf{a}$ because force and acceleration are both vectors.

EXAMPLE 8.7

A force of $(12\mathbf{i} + 3t\,\mathbf{j})\,\text{N}$, where t is the time in seconds, acts on a particle of mass 6 kg. The directions of $\mathbf{i}$ and $\mathbf{j}$ correspond to east and north respectively.

(i) Show that the acceleration is $(2\mathbf{i} + 0.5t\,\mathbf{j})\,\text{ms}^{-2}$ at time t.

(ii) Find the acceleration and magnitude of the acceleration when $t = 12$.

(iii) At what time is the acceleration directed north-east (i.e. a bearing of $045°$)?

(iv) If the particle starts with a velocity of $(2\mathbf{i} - 3\mathbf{j})\,\text{ms}^{-1}$ when $t = 0$, what will its velocity be when $t = 3$?

(v) When $t = 3$ a second **constant** force begins to act. Given that the acceleration of the particle at that time due to both forces is $4\,\text{ms}^{-2}$ due south, find the second force.

[MEI]

SOLUTION

(i) By Newton's second law the force $=$ mass $\times$ acceleration

$$(12\mathbf{i} + 3t\,\mathbf{j}) = 6\mathbf{a}$$

$$\mathbf{a} = \frac{1}{6}(12\mathbf{i} + 3t\,\mathbf{j})$$

$$\mathbf{a} = 2\mathbf{i} + 0.5t\,\mathbf{j} \quad \textcircled{1}$$

(ii) When $t = 12$

$$\mathbf{a} = 2\mathbf{i} + 6\mathbf{j}$$

magnitude of $\mathbf{a}$

$$|\mathbf{a}| = \sqrt{2^2 + 6^2}$$

The acceleration is $2\mathbf{i} + 6\mathbf{j}\,\text{ms}^{-2}$ with magnitude $6.32\,\text{ms}^{-2}$.

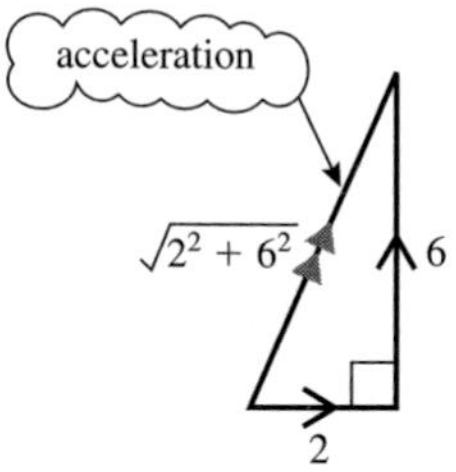

acceleration

$\sqrt{2^2 + 6^2}$ 6

2

Figure 8.13

(iii) The acceleration is north-east when its northerly component is equal to its easterly component. From $\textcircled{1}$, this is when $2 = 0.5t$ i.e. when $t = 4$.

(iv) The velocity at time t is $\int \mathbf{a}\,dt = \int (2\mathbf{i} + 0.5t\,\mathbf{j})\,dt$

$$\Rightarrow \quad \mathbf{v} = 2t\,\mathbf{i} + 0.25t^2\,\mathbf{j} + \mathbf{c}$$

The constant $\mathbf{c}$ is a vector such as $c_1\mathbf{j} + c_2\mathbf{j}$

When $t = 0$, $\mathbf{v} = 2\mathbf{i} - 3\mathbf{j}$

so

$$2\mathbf{i} - 3\mathbf{j} = 0\mathbf{i} + 0\mathbf{j} + \mathbf{c} \Rightarrow \mathbf{c} = 2\mathbf{i} - 3\mathbf{j}$$

$$\mathbf{v} = 2t\,\mathbf{i} + 0.25t^2\,\mathbf{j} + 2\mathbf{i} - 3\mathbf{j}$$

$$\mathbf{v} = (2t + 2)\,\mathbf{i} + (0.25t^2 - 3)\,\mathbf{j} \quad \textcircled{2}$$

When $t = 3$

$$\mathbf{v} = 8\mathbf{i} - 0.75\mathbf{j}$$

(v) Let the second force be **F** so the total force when $t = 3$ is
$(12\mathbf{i} + 3 \times 3\mathbf{j}) + \mathbf{F}$.

The acceleration is $-4\mathbf{j}$, so by Newton II

$$(12\mathbf{i} + 9\mathbf{j}) + \mathbf{F} = 6 \times -4\mathbf{j}$$
$$\mathbf{F} = -24\mathbf{j} - 12\mathbf{i} - 9\mathbf{j}$$
$$\mathbf{F} = -12\mathbf{i} - 33\mathbf{j}$$

The second force is $-12\mathbf{i} - 33\mathbf{j}$.

Note

You can use the same methods in three dimensions just by including a third direction **k**, for example

$\mathbf{r} = x\mathbf{i} + y\mathbf{j} + z\mathbf{k}$, $\mathbf{v} = \dot{x}\mathbf{i} + \dot{y}\mathbf{j} + \dot{z}\mathbf{k}$ and $\mathbf{a} = \ddot{x}\mathbf{i} + \ddot{y}\mathbf{j} + \ddot{z}\mathbf{k}$

and similarly in column vectors.

Figure 8.14

Historical Note

Newton's work on motion required more mathematical tools than were generally used at the time. He had to invent his own ways of thinking about continuous change and in about 1666 he produced a theory of 'fluxions' in which he imagined a quantity 'flowing' from one magnitude to another. This was the beginning of calculus. He did not publish his methods, however, and when Leibniz published his version in 1684 there was an enormous amount of controversy amongst their supporters about who was first to discover calculus. The sharing of ideas between mathematicians in Britain and the rest of Europe was hindered for a century. The contributions of both men are remembered today by their notation. Leibniz's $\dfrac{dx}{dt}$ is common and Newton's $\dot{x}$ is widely used in mechanics.

EXERCISE 8D

1 The first part of a race track is a bend. As the leading car travels round the bend its position, in metres, is modelled by:

$$\mathbf{r} = 2t^2\mathbf{i} + 8t\mathbf{j}$$

where t is in seconds.

(i) Find an expression for the velocity of the car.

(ii) Find the position of the car when $t = 0, 1, 2, 3$ and 4. Use this information to sketch the path of the car.

(iii) Find the velocity of the car when $t = 0, 1, 2, 3$ and 4. Add vectors to your sketch to represent these velocities.

(iv) Find the speed of the car as it leaves the bend at $t = 5$.

2 As a boy slides down a slide his position vector in metres at time t is

$$\begin{pmatrix} x \\ y \end{pmatrix} = \begin{pmatrix} 16 - 4t \\ 20 - 5t \end{pmatrix}$$

Find his velocity and acceleration.

3 Calculate the magnitude and direction of the acceleration of a particle that moves so that its position vector in metres is given by

$$\mathbf{r} = (8t - 2t^2)\,\mathbf{i} + (6 + 4t - t^2)\,\mathbf{j}$$

where t is the time in seconds.

4 A rocket moves with a velocity (in ms^{-1}) modelled by

$$\mathbf{v} = \tfrac{1}{10}t\,\mathbf{i} + \tfrac{1}{10}t^2\,\mathbf{j}$$

where $\mathbf{i}$ and $\mathbf{j}$ are horizontal and vertical unit vectors respectively and t is in seconds. Find

(i) an expression for its position vector relative to its starting position at time t

(ii) the displacement of the rocket after 10 s of its flight.

5 A particle is initially at rest at the origin. It experiences an acceleration given by

$$\mathbf{a} = 4t\,\mathbf{i} + (6 - 2t)\,\mathbf{j}.$$

Find expressions for the velocity and position of the particle at time t.

6 While a hockey ball is being hit it experiences an acceleration (in ms^{-2}) modelled by

$$\mathbf{a} = 1000[6t\,(t - 0.2)\,\mathbf{i} + t\,(t - 0.2)\,\mathbf{j}] \text{ for } 0 \leqslant t \leqslant 0.2 \text{ in seconds}$$

If the ball is initially at rest, find its speed when it loses contact with the stick after 0.2 s.

7 A speedboat is initially moving at 5 ms^{-1} on a bearing of 135°.

(i) Express the initial velocity as a vector in terms of $\mathbf{i}$ and $\mathbf{j}$, which are unit vectors east and north respectively.

The boat then begins to accelerate with an acceleration modelled by

$$\mathbf{a} = 0.1t\,\mathbf{i} + 0.3t\,\mathbf{j} \quad \text{in ms}^{-2}$$

(ii) Find the velocity of the boat 10 s after it begins to accelerate and its displacement over the 10 s period.

8 A girl throws a ball and, t seconds after she releases it, its position in metres relative to the point where she is standing is modelled by

$$\begin{pmatrix} x \\ y \end{pmatrix} = \begin{pmatrix} 15t \\ 2 + 16t - 5t^2 \end{pmatrix}$$

where the directions are horizontal and vertical.

(i) Find expressions for the velocity and acceleration of the ball at time t.

(ii) The vertical component of the velocity is zero when the ball is at its highest point. Find the time taken for the ball to reach this point.

(iii) When the ball hits the ground the vertical component of its position vector is zero. What is the speed of the ball when it hits the ground?

(iv) Find the equation of the trajectory of the ball as in Chapter 6, page 118.

9 The position (in metres) of a tennis ball t seconds after leaving a racquet is modelled by

$$\mathbf{r} = (20t)\,\mathbf{i} + (2 + t - 5t^2)\,\mathbf{j}$$

where $\mathbf{i}$ and $\mathbf{j}$ are horizontal and vertical unit vectors.

(i) Find the position of the tennis ball when $t = 0, 0.2, 0.4, 0.6$ and 0.8. Use these to sketch the path of the ball.

(ii) Find an expression for the velocity of the tennis ball. Use this to find the velocity of the ball when $t = 0.2$.

(iii) Find the acceleration of the ball.

(iv) Find the equation of the trajectory of the ball.

10 An owl is initially perched on a tree. It then goes for a short flight which ends when it dives on to a mouse on the ground. The position vector (in metres) of the owl t seconds into its flight is modelled by

$$\mathbf{r} = t^2(6 - t)\,\mathbf{i} + (12.5 + 4.5t^2 - t^3)\,\mathbf{j}$$

where the foot of the tree is taken to be the origin and the unit vectors $\mathbf{i}$ and $\mathbf{j}$ are horizontal and vertical.

(i) Draw a graph showing the bird's flight.

(ii) For how long (in s) is the owl in flight?

(iii) Find the speed of the owl when it catches the mouse and the angle that its flight makes with the horizontal at that instant.

(iv) Show that the owl's acceleration is never zero during the flight.

11 Ship A is 5 km due west of ship B and is travelling on a course 035° at a constant but unknown speed $v\,\mathrm{km\,h^{-1}}$. Ship B is travelling at a constant $10\,\mathrm{km\,h^{-1}}$ on a course 300°.

(i) Write the velocity of each ship in terms of unit vectors $\mathbf{i}$ and $\mathbf{j}$ with directions east and north.

(ii) Find the position vector of each ship at time t hours, relative to the starting position of ship A.

The ships are on a collision course.

(iii) Find the speed of ship A.

(iv) How much time elapses before the collision occurs?

12 A particle of mass 0.5 kg is acted on by a force, in newtons,

$$\mathbf{F} = t^2\mathbf{i} + 2t\mathbf{j}.$$

The particle is initially at rest at the origin and t is measured in seconds.

(i) Find the acceleration of the particle at time t.

(ii) Find the velocity of the particle at time t.

(iii) Find the position vector of the particle at time t.

(iv) Give all the information you can about the particle at time $t = 2$.

13 The position vector of a motorcycle of mass 150 kg on a track is modelled by

$$\mathbf{r} = 4t^2\mathbf{i} + \frac{1}{8}t\,(8-t)^2\mathbf{j} \quad 0 \leqslant t \leqslant 8$$

$$\mathbf{r} = (64t - 256)\,\mathbf{i} \qquad 8 < t \leqslant 20$$

where t is the time in seconds after the start of a race.

The vectors $\mathbf{i}$ and $\mathbf{j}$ are in directions along and perpendicular to the direction of the track as shown in the diagram. The origin is in the middle of the track. The vector $\mathbf{k}$ has direction vertically upwards.

(i) Draw a sketch to show the motorcycle's path over the first 10 s. The track is 20 m wide. Does the motorcycle leave it?

(ii) Find, in vector form, expressions for the velocity and acceleration of the motorcycle at time t for $0 \leqslant t \leqslant 20$.

(iii) Find in vector form an expression for the resultant horizontal force acting on the motorcycle during the first 8 s, in terms of t.

(iv) Why would you expect the driving force from the motorcycle's engine to be substantially greater than the component in the $\mathbf{i}$ direction of your answer to part (iii)?

When $t = 22$ s the motorcycle's velocity is given by $\mathbf{v} = 60\mathbf{i} + 6\mathbf{k}$.

(v) What has happened?

14 A small, delicate microchip which is initially at rest is to be moved by a robot arm so that it is placed **gently** on to a horizontal assembly bench. Two mathematical models have been proposed for the motion which will be programmed into the robot. In each model the unit of length is the centimetre and time is measured in seconds. The unit vectors $\mathbf{i}$ and $\mathbf{j}$ have directions which are horizontal and vertical respectively and the origin is the point O on the surface of the bench, as shown in the diagram.

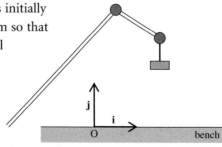

Model A for the position vector of the microchip at time t is

$$\mathbf{r}_A = 5t^2\,\mathbf{i} + (16 - 4t^2)\,\mathbf{j}, \; (t \geqslant 0).$$

(i) How far above the table is the microchip initially (i.e. when $t = 0$)?

(ii) Show that this model predicts that the microchip reaches the table after 2 s and state the horizontal distance moved in this time.

(iii) Calculate the predicted horizontal and vertical components of velocity when $t = 0$ and $t = 2$.

Model B for the position vector at time t of the microchip is

$$\mathbf{r}_B = (15t^2 - 5t^3)\,\mathbf{i} + (16 - 24t^2 + 16t^3 - 3t^4)\,\mathbf{j}, \; (t \geqslant 0).$$

(iv) Show that model B predicts the same positions for the microchip at $t = 0$ and $t = 2$ as model A.

(v) Calculate the predicted horizontal and vertical components of velocity for the microchip at $t = 0$ and $t = 2$ from model B and comment, with brief reasons, on which model you think describes the more suitable motion for the microchip.

[MEI]

KEY POINTS

Relationships between the variables describing motion

	Position	→	Velocity	→	Acceleration
			differentiate		

In one dimension s $v = \dfrac{ds}{dt}$ $a = \dfrac{dv}{dt} = \dfrac{d^2 s}{dt^2}$

In two dimensions $\mathbf{r} = x\mathbf{i} + y\mathbf{j}$ $\mathbf{v} = \dfrac{d\mathbf{r}}{dt} = \dot{x}\mathbf{i} + \dot{y}\mathbf{j}$ $\mathbf{a} = \dfrac{d\mathbf{v}}{dt} = \ddot{x}\mathbf{i} + \ddot{y}\mathbf{j}$

$= \begin{pmatrix} x \\ y \end{pmatrix}$ $= \begin{pmatrix} \dot{x} \\ \dot{y} \end{pmatrix}$ $= \begin{pmatrix} \ddot{x} \\ \ddot{y} \end{pmatrix}$

	Acceleration	→	Velocity	→	Position
			integrate		

In one dimension a $v = \int a\,dt$ $s = \int v\,dt$

In two dimensions $\mathbf{a}$ $\mathbf{v} = \int \mathbf{a}\,dt$ $\mathbf{r} = \int \mathbf{v}\,dt$

- Acceleration may be due to change in direction or change in speed or both.
- If the acceleration is *constant*:

$$\mathbf{v} = \mathbf{u} + \mathbf{a}t \quad \mathbf{r} = \mathbf{r}_0 + \mathbf{u}t + \tfrac{1}{2}\mathbf{a}t^2 \quad \mathbf{r} = \mathbf{r}_0 + \tfrac{1}{2}(\mathbf{u} + \mathbf{v})\,t$$

- Using vectors, Newton's second law is $\mathbf{F} = m\mathbf{a}$.

Answers

Chapter 1

❷ (Page 2)

(i) $+4$ **(ii)** -5

❷ (Page 3)

The marble is below the origin.

Exercise 1A (Page 4)

1 (i) $+1$ m   **(ii)** $+2.25$ m

2 (i)   3.5 m, 6 m, 6.9 m, 6 m, 3.5 m, 0 m

   (ii)   0 m, 2.5 m, 3.4 m, 2.5 m, 0 m, -3.5 m

   (iii) (a) 3.4 m   **(b)** 10.3 m

3 (i)   2 m, 0 m, -0.25 m, 0 m, 2 m, 6 m, 12 m

   (ii)

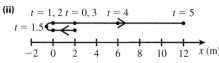

   (iii) 10 m   **(iv)** 14.5 m

4 (i)

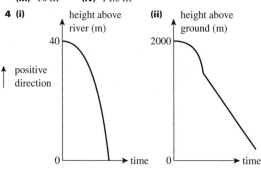

(iii)

5 (i)   The ride starts at $t = 0$. At A it changes direction and returns to pass its starting point at B continuing past to C where it changes direction again returning to its initial position at D.

   (ii)   An oscillating ride such as a swing boat.

❷ (Page 6)

10, 0, -10. The gradient represents the velocity.

❷ (Page 7)

The graph would curve where the gradient changes. Not over this period.

❷ (Page 8)

$+5\,\text{ms}^{-1}$, $0\,\text{ms}^{-1}$, $-5\,\text{ms}^{-1}$, $-6\,\text{ms}^{-1}$. The velocity decreases at a steady rate.

Exercise 1B (Page 9)

1

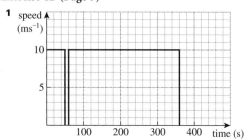

2 (i)   The person is waiting at the bus stop.

   (ii)   It is faster.

   (iii)

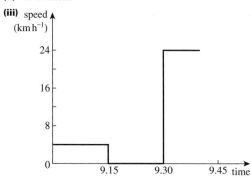

   (iv)   constant speed, infinite acceleration

3 (i)   **(a)** 2 m, 8 m   **(b)** 6 m   **(c)** 6 m

      (d) $2\,\text{ms}^{-1}$, $2\,\text{ms}^{-1}$ **(e)** $2\,\text{ms}^{-1}$ **(f)** $2\,\text{ms}^{-1}$

   (ii)   **(a)** 60 km, 0 km   **(b)** -60 km   **(c)** 60 km

      (d) $-90\,\text{km h}^{-1}$, $90\,\text{km h}^{-1}$ **(e)** $-90\,\text{km h}^{-1}$

      (f) $90\,\text{km h}^{-1}$

   (iii)   **(a)** 0 m, -10 m   **(b)** -10 m   **(c)** 50 m

      (d) OA: $10\,\text{ms}^{-1}$, $10\,\text{ms}^{-1}$; AB: $0\,\text{ms}^{-1}$, $0\,\text{ms}^{-1}$;

      BC: $-15\,\text{ms}^{-1}$, $15\,\text{ms}^{-1}$ **(e)** $-1.67\,\text{ms}^{-1}$

      (f) $8.33\,\text{ms}^{-1}$

   (iv)   **(a)** 0 km, 25 km   **(b)** 25 km   **(c)** 65 km

      (d) AB: $-10\,\text{km h}^{-1}$, $10\,\text{km h}^{-1}$; BC: $11.25\,\text{km h}^{-1}$,

      $11.25\,\text{km h}^{-1}$ **(e)** $4.167\,\text{km h}^{-1}$

      (f) $10.83\,\text{km h}^{-1}$

4 774.2 mph

❷ (Page 11)

(i) D   **(ii)** B, C, E   **(iii)** A

Exercise 1C (Page 11)

1 (i) **(a)** $+0.8\,\text{ms}^{-2}$ **(b)** $-1.4\,\text{ms}^{-2}$ **(c)** $+0.67\,\text{ms}^{-2}$
(d) 0 **(e)** $+0.5\,\text{ms}^{-2}$

(ii) acceleration

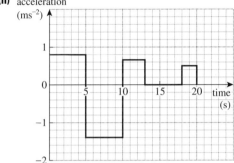

2 (i) $0\,\text{m}, -16\,\text{m}, -20\,\text{m}, 0\,\text{m}, 56\,\text{m}.$

(ii)

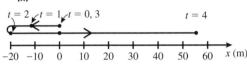

(iii)

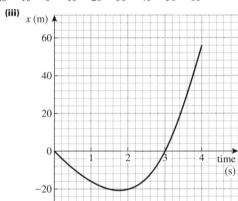

(iv) after $0\,\text{s}$ (negative direction) and $3\,\text{s}$ (positive direction).

3 (i) $20\,\text{mph}$ **(ii)** $22.3\,\text{mph}$

4 (i) **(a)** $36\,\text{mph}$ **(b)** $58.5\,\text{mph}$ **(c)** $30\,\text{mph}$

(ii) The average speed is not equal to the mean value of the two speeds unless the same time is spent at the two speeds. In this case the ratio of distances must be 3:1.

5 (i) velocity (ms^{-1})

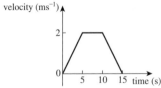

(ii) velocity (ms^{-1})

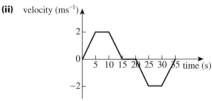

(iii) $+0.4\,\text{ms}^{-2}, 0\,\text{ms}^{-2}, -0.4\,\text{ms}^{-2}, 0\,\text{ms}^{-2}, -0.4\,\text{ms}^{-2},$
$0\,\text{ms}^{-2}, +0.4\,\text{ms}^{-2}$

(iv) acceleration

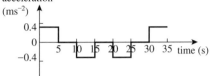

6 (i) $s\,(\text{m})$

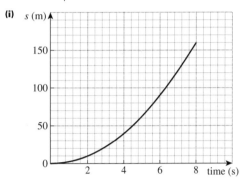

(ii) (a)

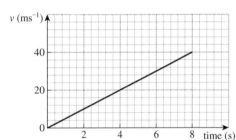

(b)

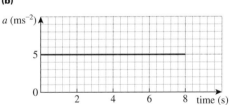

❓ (Page 13)

(i) 5 **(ii)** 20 **(iii)** 45 They are the same.

❓ (Page 13)

It represents the displacement.

❓ (Page 15)

Approx $460\,\text{m}$

❓ (Page 15)

No, so long as the lengths of the parallel sides are unchanged the trapezium has the same area.

Exercise 1D (Page 16)

1 Car A **(i)** $0.4\,\text{ms}^{-2}$, $0\,\text{ms}^{-2}$, $3\,\text{ms}^{-2}$

(ii) $62.5\,\text{m}$ **(iii)** $4.17\,\text{ms}^{-1}$

Car B **(i)** $-1.375\,\text{ms}^{-2}$, $-0.5\,\text{ms}^{-2}$, $0\,\text{ms}^{-2}$, $2\,\text{ms}^{-2}$

(ii) $108\,\text{m}$ **(iii)** $3.6\,\text{ms}^{-1}$

2 (i) Enters motorway at $10\,\text{ms}^{-1}$, accelerates to $30\,\text{ms}^{-1}$ and maintains this speed for about $150\,\text{s}$. Slows down to stop after a total of $400\,\text{s}$.

(ii) Approx. $0.4\,\text{ms}^{-2}$, $-0.4\,\text{ms}^{-2}$

(iii) Approx. $9.6\,\text{km}$, $24\,\text{ms}^{-1}$

3 (i)

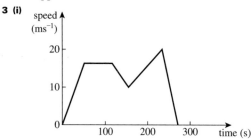

(ii) $3562.5\,\text{m}$

4 (i)

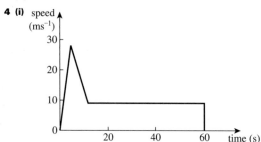

(ii) $558\,\text{m}$

5 (i)

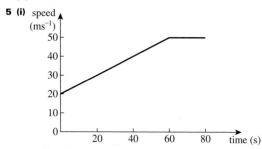

(ii) after $60\,\text{s}$

(iii) $6.6\,\text{km}$

(iv) $v = 20 + 0.5t$ for $0 \leqslant t \leqslant 60$,

$v = 50$ for $t \geqslant 60$

6 (i)

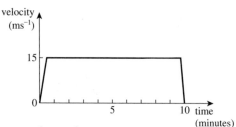

(ii) $15\,\text{ms}^{-1}$, $1\,\text{ms}^{-2}$, $8.66\,\text{km}$

7 BC: constant deceleration, CD: stationary, DE: constant acceleration

(ii) $-0.5\,\text{ms}^{-2}$, $2500\,\text{m}$

(iii) $0.2\,\text{ms}^{-2}$, $6250\,\text{m}$

(iv) $325\,\text{s}$

(v)

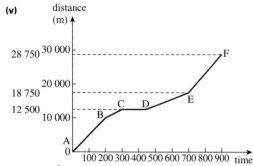

8 (i) $10\,\text{ms}^{-1}$, $0.7\,\text{s}$

(ii)

velocity (ms^{-1})

(iii) $6.25\,\text{ms}^{-2}$

(iv) $33.9\,\text{m}$

Chapter 2

? (Page 20)

See text which follows.

? (Page 21)

It might be reasonable for sections of motorway depending on the traffic.

? (Page 26)

$u = 13.3\,\text{ms}^{-1}$, $v = 26.7\,\text{ms}^{-1}$, $t = 5$, $s = 100\,\text{m}$

? (Page 27)

$s = \frac{1}{2}(2u + at) \times t$

$s = (u + \frac{1}{2}at) \times t$

$s = ut + \frac{1}{2}at^2$

? (Page 27)

$s = ut + \frac{1}{2}at^2$

$s = (v - at) \times t + \frac{1}{2}at^2$

$s = vt - at^2 + \frac{1}{2}at^2$

$s = vt - \frac{1}{2}at^2$

Exercise 2A (Page 28)

1 (i) 22   (ii) 120   (iii) 0   (iv) −10

2 (i) $v^2 = u^2 + 2as$

   (ii) $v = u + at$

   (iii) $s = ut + \frac{1}{2}at^2$

   (iv) $s = \dfrac{(u + v)}{2} \times t$

   (v) $v^2 = u^2 + 2as$

   (vi) $s = ut + \frac{1}{2}at^2$

   (vii) $v^2 = u^2 + 2as$

   (viii) $s = vt - \frac{1}{2}at^2$

3 (i) $9.8\,\mathrm{ms^{-1}}$, $98\,\mathrm{ms^{-1}}$ (ii) $4.9\,\mathrm{m}$, $490\,\mathrm{m}$ (iii) 2 s.
Speed and distance after 10 s, both over-estimates.

4 $2.08\,\mathrm{ms^{-2}}$, 150 m. Assume constant acceleration.

5 $4.5\,\mathrm{ms^{-2}}$, 9 m

6 $-8\,\mathrm{ms^{-2}}$, 3 s

7 (i) $s = 16t - 4t^2$, $v = 16 - 8t$

   (ii)   (a) 2 s   (b) 4 s

   (iii)

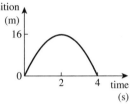

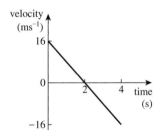

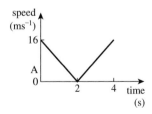

❓ (Page 30)

$u = -15.0$, No

❓ (Page 33)

$x = 5t^2$ and $1.25 - x = 5t - 5t^2$ so $1.25 = 5t$ as before.

❓ (Page 33)

Because the velocity is not constant.

Exercise 2B (Page 33)

1 604.9 s, 9037 m or 9.04 km

2 (i) $v = 2 + 0.4t$

   (ii) $s = 2t + 0.2t^2$

   (iii) $18\,\mathrm{ms^{-1}}$

3 No, he is 10 m behind when Sabina finishes.

4 (i) $h = 2t - 4.9t^2 + 4$

   (ii) 1.13 s   (iii) $9.08\,\mathrm{ms^{-1}}$ (iv) t greater, v less

5 (i) $12\,\mathrm{ms^{-1}}$ (ii) 8.45 m   (iii) $13\,\mathrm{ms^{-1}}$ (iv) 5.41 m

   (v) under-estimate

6 (i)

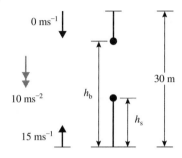

   (ii) $h_s = 15t - 5t^2$ (iii) $h_b = 30 - 5t^2$

   (iv) $t = 2\,\mathrm{s}$ (v) 10 m

7 (i) $5.4\,\mathrm{ms^{-1}}$ (ii) $-4.4\,\mathrm{ms^{-1}}$ (iii) $1\,\mathrm{ms^{-1}}$ increase

   (iv) $9\,\mathrm{ms^{-1}}$ (v) too fast

8 3 m

9 43.75 m

10 (ii) $u + 5a = 15.15$ (iii) 14.4

   (iv) No, distance at constant acceleration is 166.5 m.

Chapter 3

❓ (Page 40)

The reaction between person and chair acts on chair.
The person's weight acts on the person only.

❓ (Page 42)

Vertically up.

Exercise 3A (Page 43)

In these diagrams, W represents a weight, N a normal reaction with another surface, F a friction force, R air resistance and P another force.

1

2

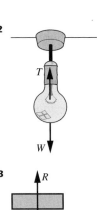

3

4

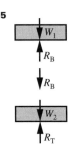

5

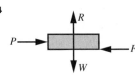

6

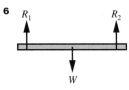

7

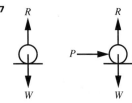

8

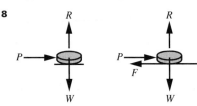

9

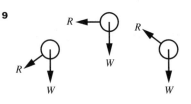

10

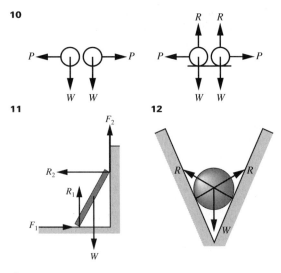

11

12

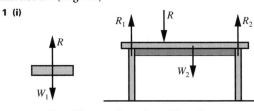

❓ (Page 44)

To provide forces when the velocity changes.

❓ (Page 45)

The friction force was insufficient to enable his car to change direction at the bend.

Exercise 3B (Page 46)

1 (i)

(ii) (a) $R = W_1$ **(b)** $R_1 + R_2 = W_2 + R$
2 (i) $R = W, 0$ **(ii)** $W > R, W - R$ down
   (iii) $R > W, R - W$ up
3 (i) No   **(ii)** Yes   **(iii)** Yes   **(iv)** No   **(v)** Yes
   (vi) Yes   **(vii)** Yes   **(viii)** No
4 Forces are required to give passengers the same acceleration as the car.
   (i) A seat belt provides a backwards force.
   (ii) The seat provides a forwards force on the body and the head rest is required to make the head move with the body.

❓ (Page 48)

Figure 3.18

Exercise 3C (Page 52)

1 (i) 147 N   **(ii)** 11 760 N = 11.76 kN   **(iii)** 0.49 N
2 (i) 61.2 kg   **(ii)** 1120 kg = 1.12 tonne
3 (i) 637 N   **(ii)** 637 N
4 112 N

5 (i) Both hit the ground together.

(ii) The balls take longer to hit the ground on the moon, but still do so together.

6 Answers for 60 kg **(ii)** 588 N **(iii)** 96 N

(iv) its mass is 4 kg

? (Page 52)

No. Scales which measure by balancing an object against fixed masses (weights).

Exercise 3D (Page 54)

In these diagrams, mg *represents a weight,* N *a normal reaction with another surface,* F *a friction force,* R *air resistance,* T *a tension or thrust,* D *a driving force and* P *another force.*

1 (i)

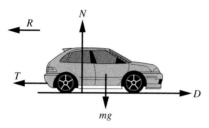

(ii)

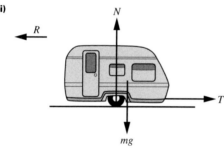

(iii)

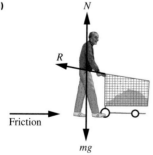

(iv) (a)

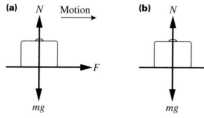

(v)

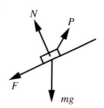

2 (i) Weight $5g$ down and reaction ($=5g$) up.

(ii) Weight $5g$ down, reaction with box above ($=45g$ down) and reaction with ground ($=50g$ up).

3 (i) $F_1 = 10$ **(ii)** $15 - F_2$ N

4 (i) towards the left

(ii)

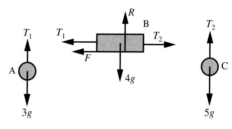

(iii) $3g$ N, $5g$ N

(iv) $2g$ N

(v) $T_1 - 3g \uparrow$, $T_2 - T_1 - F \rightarrow$, $5g - T_2 \downarrow$

5 All forces are in newtons

(i) greater

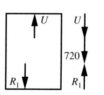

(ii) less

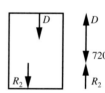

(iii) greater

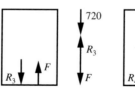

(iv) less

6 (i) 2400 N

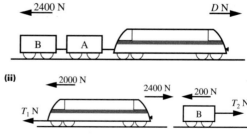

(ii)

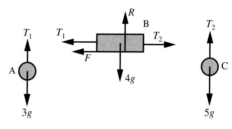

(iii) $T_1 = 400$

(iv) $T_2 = 200$

Chapter 4

❓ (Page 58)

The pointer moves up and down as the force on the spring varies. Your weight would seem to change as the speed of the lift changed. You feel the reaction force between your hand and the book which varies as you move the book up and down.

Exercise 4A (Page 59)

1 (i) 800 N **(ii)** 88 500 N **(iii)** 0.0225 N
(iv) 840 000 N **(v)** 8×10^{-20} N **(vi)** 548.8 N
(vii) 8.75×10^{-5} N **(viii)** 10^{30} N
2 (i) 200 kg **(ii)** 50 kg **(iii)** 10 000 kg **(iv)** 1.02 kg
3 (i) 7.6 N **(ii)** 7.84 N

❓ (Page 61)

There is a resultant downwards force because the weight is greater than the tension.

Exercise 4B (Page 63)

1 (i) 0.5 ms^{-2} **(ii)** 25 m
2 (i) 1.67 ms^{-2} **(ii)** 16.2 s
3 (i) 325 N **(ii)** 1764 N
4 (i) 13 N **(ii)** 90 m **(iii)** 13 N
5 (i)

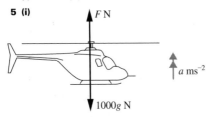

(ii) 11 300 N
6 (i) $400 - 250 = 12\,000a$, $a = 0.0125$ ms^{-2}
(ii) 0.5 ms^{-1}, 40 s
(iii) (a) 15 s **(b)** 13.75 m **(c)** 55 s
7 (i) 60 ms^{-1} **(ii)** continues at 60 ms^{-1}
(iii) 1.25 N **(iv)** the first by 655 km
8 (i) 6895 N, 6860 N, 6790 N, 1960 N
(ii) 815 kg **(iii)** max $T < 9016$ N
9 (i) 7.84 ms^{-2} **(ii)** 13.7 ms^{-1} is just over 30 mph

❓ (Page 65)

Your own weight acts on you and the tensions in the ropes with which you have contact, the other person's weight acts on them. The tension forces acting at the ends of the rope AB are equal and opposite. The accelerations of A and B are equal because they must always travel the same distance in each interval of time assuming the rope does not stretch.

❓ (Page 66)

The tension in the rope joining A and B must be greater than B's weight because there must be a resultant force on B to produce an acceleration.

❓ (Page 69)

Using $v = u + at$ with $u = 0$ and maximum $a = 6.2$, the speed after 1 second would be 6.2 ms^{-1} or about 14 mph. Under the circumstances, a careful driver is unlikely to accelerate at this rate.

Alvin and his snowmobile and Bernard are two particles each moving in a straight line, otherwise Bernard could swing from side to side; contact between the ice and the rope is smooth, otherwise, the tensions acting on Alvin and Bernard are different; the rope is light, otherwise its tension would be affected by its weight; the rope is of constant length, otherwise the accelerations would not be equal; there is no air resistance, otherwise the equations of motion would involve a force to allow for it.

Exercise 4C (Page 71)

1 (i)

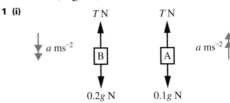

(ii) $T - 0.1g = 0.1a$, $0.2g - T = 0.2a$
(iii) 3.27 ms^{-2}, 1.31 N **(iv)** 1.11 s
2 (i)

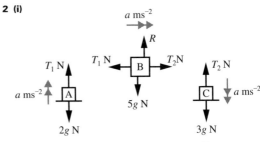

(ii) $T_1 - 2g = 2a$, $T_2 - T_1 = 5a$, $3g - T_2 = 3a$
(iii) 1 ms^{-1}, 22 N, 27 N **(iv)** 5 N
3 (ii) 750 N **(iii)** tension, 44 N **(iv)** 170 N
4 (i) 0.625 ms^{-2}
(ii) 25 000 N

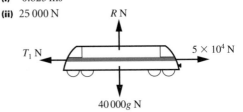

(iii) 12 500 N **(iv)** reduced to 10 000 N

5 (i) $0.25\,\text{ms}^{-2}$

(ii) 9000 N, 6000 N

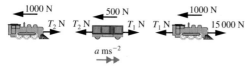

(iii) $0.25\,\text{ms}^{-2}$, tension 1500 N, thrust 1500 N
The second engine is now pushing rather than
pulling back on the truck.

6 (i)

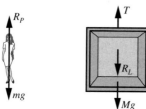

(ii) $R_P = R_L = 50g$, $T = 500g$

(iii) $T = 5300$, $R_P = R_L = 530$

7 (i)

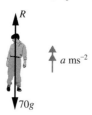

(ii) $1\,\text{ms}^{-2}$

(iii) stationary for 2 s, accelerates at $1\,\text{ms}^{-2}$ for 2 s,
constant speed for 5 s, decelerates at $2\,\text{ms}^{-2}$ for
1 s, stationary for 2 s.

(iv)

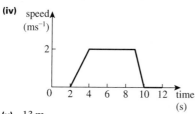

(v) 13 m

8 (i)

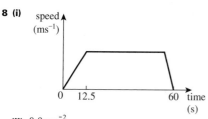

(ii) $0.8\,\text{ms}^{-2}$

(iii) 40 s, $1.33\,\text{ms}^{-2}$

(iv) 66.4 kN   **(v)** 90 kg

❓ (Page 74)

The air resistance seems to affect them more.

❓ (Page 75)

Sky divers and flying squirrels maximise air resistance
by presenting a larger surface area in the direction of
motion. Cyclists minimise air resistance by reducing
the area.

❓ (Page 76)

Yes

❓ (Page 76)

Air resistance depends on velocity through the air. The
velocities of a pair of cards in the experiment do not
differ very much over such small heights.

Chapter 5

❓ (Page 78)

See text which follows.

❓ (Page 80)

If the bird flies, say, 5 cm N, the wind would blow it
12 cm E and its resultant displacement would be 13 cm
along DF. This would occur in any small interval of
time.

Exercise 5A (Page 82)

1 $c = -b$, $d = -a$, $e = a$, $f = 1.5a$, $g = -2b$, $|a| = |b|$

2 3 m, $+50°$; 3 m, $-40°$; 1 m, horizontal

3 $32\,\text{ms}^{-1}$ and $28\,\text{ms}^{-1}$

4 (i) $8\,\text{ms}^{-1}$ downstream

   (ii) $2\,\text{ms}^{-1}$ upstream

   (iii) $3\,\text{ms}^{-1}$ downstream

5

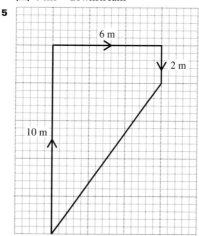

10 m at 53° to the horizontal

6 (i) $\mathbf{a} + \mathbf{b} = \mathbf{b} + \mathbf{a}$

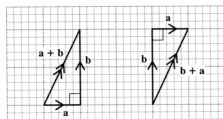

(ii) $\mathbf{a} + \mathbf{b} + \mathbf{c} = \mathbf{c} + \mathbf{a} + \mathbf{b}$

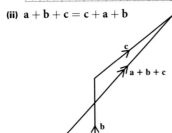

7 (i) $\overrightarrow{AB} = 2\overrightarrow{DC}$ **(ii)** $\overrightarrow{AB} = -2\overrightarrow{CD}$
 (iii) $\overrightarrow{EA} = 0.7\overrightarrow{CB}$ **(iv)** $\overrightarrow{AE} = -0.7\overrightarrow{CB}$
8 $\overrightarrow{AB} + \overrightarrow{BC} = \overrightarrow{AC},\ \overrightarrow{AC} + \overrightarrow{CD} = \overrightarrow{AD},$
 $\overrightarrow{AD} + \overrightarrow{DA} = 0$
9 (i) $\mathbf{p} + \mathbf{q}$ **(ii)** $\frac{1}{3}\mathbf{q}$ **(iii)** $\mathbf{p} + \frac{1}{3}\mathbf{q}$ **(iv)** $\frac{1}{2}\mathbf{p} + \frac{1}{6}\mathbf{q}$
10 (i) $\mathbf{a}$ **(ii)** $\mathbf{c} + \mathbf{a}$ **(iii)** $-\mathbf{a} + \mathbf{c}$ **(iv)** $\mathbf{a} - \mathbf{c}$
 (v) $-\mathbf{a} - \mathbf{c}$ **(vi)** $\frac{1}{2}\mathbf{c}$ **(vii)** $\mathbf{a} + \frac{1}{2}\mathbf{c}$ **(viii)** $\frac{1}{2}\mathbf{c} - \mathbf{a}$
11 (i) $-\mathbf{p}$ **(ii)** $\mathbf{q} - \mathbf{p}$ **(iii)** $\mathbf{q} - \mathbf{p}$ **(iv)** $\mathbf{q} - 2\mathbf{p}$
 (v) $2\mathbf{q} - 3\mathbf{p}$ **(vi)** $\mathbf{q} - 2\mathbf{p}$

? (Page 85)

$\sqrt{18}\mathbf{i} + \sqrt{2}\mathbf{j}$

? (Page 86)

-6

? (Page 87)

$\overrightarrow{AO} + \overrightarrow{OB} = -\mathbf{a} + \mathbf{b} = \mathbf{b} - \mathbf{a}$

Exercise 5B (Page 88)
1 (i) 0 m E, 2 m N **(ii)** –6 m E, 0 m N
 (iii) 6 m E, 4 m N
2 $\mathbf{a} = -2\mathbf{i},\ \mathbf{b} = \mathbf{j},\ \mathbf{c} = -3\mathbf{i},\ \mathbf{d} = 3\mathbf{j},\ \mathbf{e} = 2\mathbf{i},\ \mathbf{f} = \mathbf{i} + \mathbf{j},$
 $\mathbf{g} = -2\mathbf{i} - \mathbf{j},\ \mathbf{h} = \mathbf{i} - 2\mathbf{j},\ \mathbf{k} = \mathbf{i} - \mathbf{j}$
3 $(4, -11)$
4 (i) $9\mathbf{i} - 3\mathbf{j}$ **(ii)** $4\mathbf{i} - 2\mathbf{j}$ **(iii)** $-7\mathbf{i} + 2\mathbf{j}$
 (iv) $4\mathbf{i} - 6\mathbf{j}$ **(v)** $-3\mathbf{i} + 11\mathbf{j}$ **(vi)** $-8\mathbf{i} + 8\mathbf{j}$
5 (i) $\binom{2}{1}$ **(ii)** $\binom{-10}{-24}$ **(iii)** $\binom{0}{-2}$ **(iv)** $\binom{-3}{22}$
6 (i) $\mathbf{i} + 2\mathbf{j}, 5\mathbf{i} + \mathbf{j}, 7\mathbf{i} + 8\mathbf{j}$
 (ii) $4\mathbf{i} - \mathbf{j}, 2\mathbf{i} + 7\mathbf{j}, -6\mathbf{i} - 6\mathbf{j}$

(iii)

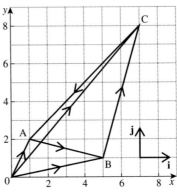

7 (i) $-3\mathbf{j}, 2\mathbf{i} + 5\mathbf{j}, 3\mathbf{i} + 9\mathbf{j}$ **(ii)** $2\mathbf{i} + 8\mathbf{j}, \mathbf{i} + 4\mathbf{j}$
 (iii) BC is parallel to AB
8 (i) $d = 9$
 (ii) BC is equal and parallel to AD so ABCD is a
 parallelogram.
9 $9\mathbf{i} + 6\mathbf{j}$
10 (i) $\frac{1}{2}(8\mathbf{i} - 11\mathbf{j})$
 (ii) $\frac{19}{3}\mathbf{i} - 16\mathbf{j}$
11 $-4.5, 10.5$

Exercise 5C (Page 92)
1 (i) 3.61 at $56.3°$ **(ii)** 13 at $-67.4°$
 (iii) 4.12 at $166°$ **(iv)** 6.71 at $-117°$
2 (i) 10 at $-53°$ **(ii)** 4 at $180°$ **(iii)** 2.24 at $-117°$
3 $30\mathbf{i} + 30\mathbf{j}, 42.4$ at $45°$
4 $-\mathbf{i} + 2\mathbf{j}, 2.24$ at $117°$
5 (ii) (a) $0.8\mathbf{i} + 0.6\mathbf{j}$ **(b)** $0.71\mathbf{i} - 0.71\mathbf{j}$
6 (ii) (a) $\frac{2}{7}\mathbf{i} - \frac{6}{7}\mathbf{j} + \frac{3}{7}\mathbf{k}$ **(b)** $0.577(\mathbf{i} + \mathbf{j} + \mathbf{k})$
7 (i) $\frac{5}{13}\mathbf{i} - \frac{12}{13}\mathbf{j}$ **(ii)** $\mathbf{F} = 15\mathbf{i} - 36\mathbf{j}$
8 $14\mathbf{i} + 48\mathbf{j}$
9 (i) $-3\mathbf{i} + \mathbf{j} + 2\mathbf{k}, 5\mathbf{i} - 4\mathbf{j} + 3\mathbf{k}, -8\mathbf{i} + 5\mathbf{j} - \mathbf{k}$
 (ii) 2.45, 6.78, 2.83

Exercise 5D (Page 95)
1 (i) $5.64\mathbf{i} + 2.05\mathbf{j}$ **(ii)** $-5.36\mathbf{i} + 4.50\mathbf{j}$
 (iii) $1.93\mathbf{i} - 2.30\mathbf{j}$ **(iv)** $-1.45\mathbf{i} - 2.51\mathbf{j}$
2 (i) $113\mathbf{i} + 65\mathbf{j}$ **(ii)** $192\mathbf{i} - 161\mathbf{j}$

(iii) $-200\mathbf{i} - 346\mathbf{j}$ **(iv)** $-43\mathbf{i} + 25\mathbf{j}$

3 $2.83\mathbf{i} + 2.83\mathbf{j}$, $3\mathbf{i}$, $6.48\,\text{km h}^{-1}$ at $064°$

4 (i)

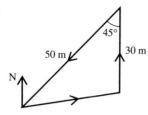

(ii) $\begin{pmatrix} 0 \\ 30 \end{pmatrix}$, $\begin{pmatrix} -35.4 \\ -35.4 \end{pmatrix}$ (iii) $081°$

5 (i) (a) $\mathbf{p} = -0.92\mathbf{i} + 2.54\mathbf{j}$, $\mathbf{q} = 2.30\mathbf{i} + 1.93\mathbf{j}$,
$\mathbf{r} = 1.7\mathbf{i} - 2.94\mathbf{j}$, $\mathbf{s} = -2.42\mathbf{i} - 1.4\mathbf{j}$
(b) $\mathbf{p} + \mathbf{q} + \mathbf{r} + \mathbf{s} = 0.66\mathbf{i} + 0.13\mathbf{j}$
(ii) (a) $\mathbf{t} = 1.35\mathbf{i} + 2.34\mathbf{j}$, $\mathbf{u} = 2.68\mathbf{i} - 1.55\mathbf{j}$,
$\mathbf{v} = -0.35\mathbf{i} - 1.97\mathbf{j}$, $\mathbf{w} = -2\mathbf{i}$
(b) $\mathbf{t} + \mathbf{u} + \mathbf{v} + \mathbf{w} = 1.683\mathbf{i} + 1.18\mathbf{j}$

6 (i) $1.45\,\text{km}$ at $046°$ (ii) $0\,\text{m}$

7 $64.3\mathbf{i} + 76.6\mathbf{j}$, $-153.2\mathbf{i} + 128.6\mathbf{j}$, $-88.9\mathbf{i} + 205.2\mathbf{j}$

8 (i)

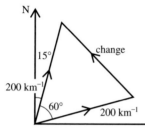

(ii) $-141\mathbf{i} + 141\mathbf{j}$

9 $5\mathbf{i} - \mathbf{j}$, $0.87\mathbf{i} + 3.92\mathbf{j}$, $-3.21\mathbf{i} - 4.83\mathbf{j}$, $-6\mathbf{j}$

10 $5544\,\text{km}$, $051°$

11 $079°$, $5.1\,\text{km}$

12 (i) $a\mathbf{i}$, $-b\cos\theta\mathbf{i} + b\sin\theta\mathbf{j}$
(ii) $\mathbf{a} + \mathbf{b} = (a - b\cos\theta)\mathbf{i} + b\sin\theta\mathbf{j}$
$|\mathbf{a} + \mathbf{b}|^2 = (a - b\cos\theta)^2 + b^2\sin^2\theta$
(iii) $c^2 = a^2 + b^2 - 2ab\cos\theta$

Exercise 5E (Page 98)

1 4.42 knots

2 $329.5\,\text{km h}^{-1}$, $355.1°$

3 (i) 11.8 knots, $283°$ (ii) $255°$

4 (i) $60°$ (ii) $7.7\,\text{s}$

5 $175.7°$, $1\,\text{h}\,5\,\text{m}$

6 (i) $3\,\text{pm}$ (ii) $150\,\text{km}$ (iii) $97.6°$

7 (i) $092°$ from A to B, $268°$ from B to A   (ii) $5.3\,\text{km}$

8 $50\,\text{ms}^{-1}$, $150\,\text{ms}^{-1}$

9 (i) $18.2\,\text{ms}^{-1}$ (ii) $0.353\,\text{ms}^{-2}$

Chapter 6

❷ (Page 103)

(i) the vertical component of velocity is zero
(ii) $y = 0$

---

❷ (Page 104)

1 Yes for a parabolic path. $u_y - gt = 0$ when $t = \dfrac{u_y}{g}$
and $u_y t - \frac{1}{2}gt^2 = 0$ when $t = 2\dfrac{u_y}{g}$.

2 The balls and the bullet can be modelled as projectiles when there is no spin or wind and air resistance is negligible. Also a rocket with no power. The air affects the motion of the others.

Exercise 6A (Page 105)

1 (i) (a)

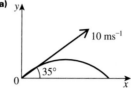

(b) $u_x = 8.2$
$u_y = 5.7$
(c) $v_x = 8.2$
$v_y = 5.7 - 9.8t$
(d) $x = 8.2t$
$y = 5.7t - 4.9t^2$

(ii) (a)

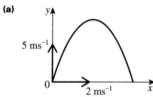

(b) $u_x = 2$
$u_y = 5$
(c) $v_x = 2$
$v_y = 5 - 9.8t$
(d) $x = 2t$
$y = 5t - 4.9t^2$

(iii) (a)

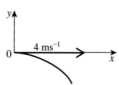

(b) $u_x = 4$
$u_y = 0$
(c) $v_x = 4$
$v_y = -9.8t$
(d) $x = 4t$
$y = -4.9t^2$

(iv) (a)

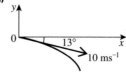

(b) $u_x = 9.7$
$u_y = -2.2$

(c) $v_x = 9.7$
$v_y = -2.2 - 9.8t$

(d) $x = 9.7t$
$y = -2.2t - 4.9t^2$

(v) (a)

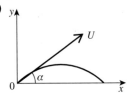

(b) $u_x = U \cos \alpha$
$u_y = U \sin \alpha$

(c) $v_x = U \cos \alpha$
$v_y = U \sin \alpha - gt$

(d) $x = Ut \cos \alpha$
$y = Ut \sin \alpha - \frac{1}{2}gt^2$

(vi) (a)

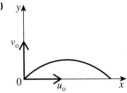

(b) $u_x = u_o$
$u_y = v_o$

(c) $v_x = u_o$
$v_y = v_o - gt$

(d) $x = u_o t$
$y = v_o t - \frac{1}{2}gt^2$

2 (i) (a) 1.5 s **(b)** 11 m

  (ii) (a) 0.5 s **(b)** 1.25 m

3 (i) (a) 4 s **(b)** 80 m

  (ii) 0.88 s **(b)** 2.21 m

Exercise 6B (Page 108)

1 (i) (a)

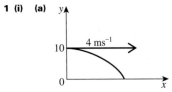

(b) $\begin{pmatrix} 4 \\ -9.8t \end{pmatrix}$ **(c)** $\begin{pmatrix} 4t \\ 10 - 4.9t^2 \end{pmatrix}$

(ii) (a)

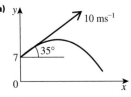

(b) $\begin{pmatrix} 8.2 \\ 5.7 - 9.8t \end{pmatrix}$ **(c)** $\begin{pmatrix} 8.2t \\ 7 + 5.7t - 4.9t^2 \end{pmatrix}$

(iii) (a)

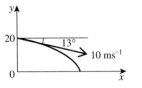

(b) $\begin{pmatrix} 9.7 \\ -2.2 - 9.8t \end{pmatrix}$ **(c)** $\begin{pmatrix} 9.7t \\ 20 - 2.2t - 4.9t^2 \end{pmatrix}$

(iv) (a)

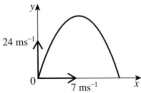

(b) $\begin{pmatrix} 7 \\ 24 - 9.8t \end{pmatrix}$ **(c)** $\begin{pmatrix} 7t \\ 24t - 4.9t^2 \end{pmatrix}$

(v) (a)

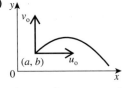

(b) $\begin{pmatrix} u_o \\ v_o - gt \end{pmatrix}$ **(c)** $\begin{pmatrix} a + u_o t \\ b + v_o t - \frac{1}{2}gt^2 \end{pmatrix}$

2 (i) (a) 1.5 s **(b)** 26 m

  (ii) (a) 0.31 s **(b)** 10.46 m

3 (i) 2.86 m **(ii)** 2.86 m

  (iii)

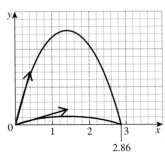

2.86

❓ (Page 113)

They land together because *u*, *s* and *a* in the vertical direction are the same for both.

Exercise 6C (Page 113)

1 (i) 17.3, 10 ms^{-1} **(ii)** 0, -10 ms^{-2}

  (iii) 35.3 m **(iv)** 1 s **(v)** 5 m

2 (i) 41, 28.7 ms^{-1}

  (ii)

t	0	1	2	3	4	5	6
x	0	41	82	123	164	205	246
y	0	24	38	42	36	21	-4.3

(iii)

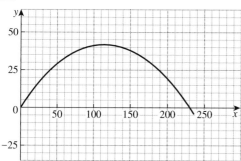

(iv) 42 m, 239.7 m

(v) The ball is a particle, no spin, no air resistance so vertical acceleration $= g$

3 **(i)** 17.2, 8 ms^{-1} **(ii)** 1.64 s **(iii)** 28.2 m
(iv) 0.82 s **(v)** 3.3 m **(vi)** 2.72 m, No

4 **(i)** 10.3, 14.7 ms^{-1} **(ii)** 2.91 s
(iii) Into the goal **(iv)** No

5 45.2 s **(ii)** 13.6 km **(iii)** 535 ms^{-1} **(iv)** 56°

6 **(i)** 0.47 s **(ii)** 0.64 s **(iii)** 25.5 ms^{-1}
(iv) 28.8 ms^{-1}

7 **(i)** Yes, the range is 70.4 m. **(ii)** 32.7 ms^{-1}

8 **(i)** 2.02 s **(ii)** No, height $= 0.2$ m
(iii) 21.57 ms^{-1}
(iv) Spin causes the ball to rise more.

9 **(i)** **(a)** 34.6 m **(b)** 39.4 m **(c)** 40 m **(d)** 39.4 m
(e) 34.6 m
(ii) $80 \sin \alpha \cos \alpha = 80 \cos (90 - \alpha) \sin (90 - \alpha)$
(iii) 57.9
(iv) $+30$ cm, -31 cm; lower angle slightly more accurate.

10 **(i)** 26.1 ms^{-1} **(ii)** 27.35 ms^{-1}
(iii) $26.15 < u < 26.88$

11 25.5 m

12 **(i)** 1.74 s **(ii)** 3.5 m, Juliet's window
(iii) 9.12 ms^{-1}

13 **(i)** 3.2 m, vertical component of velocity is always $\leqslant 8$ ms^{-1}
(ii) 5.5 m **(iii)** 52°

14 **(i)** 2 s **(ii)** $y_1 = y_2 = 19.6 - 4.9t^2$ at all times
(iii) 13.9 ms^{-1} **(iv)** $v < 10$

15 **(i)** **(a)** 14.1 ms^{-1}, 45° **(b)** 30 ms^{-1}, horizontal
(ii) $10(1 + 2k)t$, $10(1 - k)t - 5t^2$
(iii) $2(1 - k)$ s **(iv)** 22.5 m

? (Page 118)

$20\mathbf{i} + 30\mathbf{j}$ ms^{-1}, $(0, 6)$, 10 ms^{-2}.

Exercise 6D (Page 119)

1 **(i)** $y = \frac{5}{16}x^2$

(ii) $y = 6 + 0.4x - 0.2x^2$
(iii) $y = -14 + 14x - 5x^2$
(iv) $y = 5.8 + 2.4x - 0.2x^2$
(v) $y = 2x - \dfrac{gx^2}{2u^2}$

2 **(i)** $x = 40t$
(iii)

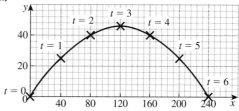

3 **(i)** $y = \frac{3}{4}x - \frac{1}{320}x^2$
(ii) Air resistance would reduce x.

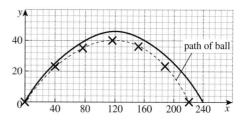

(iii) Yes, horizontal acceleration $= -0.5$ ms^{-2}

? (Page 121)

5.1 m

Exercise 6E (Page 123)

1 **(i)** $(21.2t, 21.2t - 5t^2)$ **(iii)** 8.9 **(iv)** 29.7 or 61.2

2 **(ii)** 6.9 m

3 **(ii)** 24.1 m **(iii)** Yes, $y = 2.4$ m

4 **(i)** $y = 1 + 0.7x - \dfrac{7.45x^2}{u^2}$
(ii) $u > 7.8$ ms^{-1}
(iii) $u < 8.5$ ms^{-1}

5 **(ii)** 63°, 83° **(iii)** **(a)** Yes **(b)** No

6 **(i)** $y = 2 + x \tan \alpha - \dfrac{x^2}{320}(1 + \tan^2\alpha)$
(ii) 87°, −3.1° **(iii)** **(a)** No **(b)** Yes

7 **(i)** $x = 25t \cos \alpha$, $y = 1 + 25t \sin \alpha - 5t^2$
(ii) $y = 1 + x \tan \alpha - \dfrac{x^2}{125}(1 + \tan^2 \alpha)$
(iii) **(a)** No **(b)** Yes **(iv)** 13°, 83°

8 **(i)** $x = 15t \cos \alpha$
(iv) No, 45° gives the maximum range level with the point of projection. In this case 44° gives greater X.

9 **(ii)** 1.8, 0.6 **(iii)** 22.5 m when $\tan \alpha = 1.2$
(iv) 5.7 m

10 **(i)** $y = \dfrac{4x}{3} - \dfrac{x^2}{180}$
(ii) $\dfrac{y}{x} = \tan \beta = \dfrac{1}{2}$
(iii) $(150, 75)$ **(iv)** 170 m **(v)** 140 m **(vi)** No

❓ (Page 127)

The projectile is a particle and there is no air resistance or wind. The particle is projected from the origin.

Experiment (Page 128)

1 $g \sin \theta$ where θ is the angle between the table and the horizontal
2 not according to the simplest model
3 Yes, at angles α and $(90° - \alpha)$
4 45°
5 a parabola

Chapter 7

❓ (Page 130)

Yes if the cable makes small angles with the horizontal. See the experiment on page 135.

❓ (Page 131)

Parallel to the slope.

❓ (Page 132)

Start with AB and BC. Then draw a line in the right direction for CD and another perpendicular line through A. These lines meet at D.

❓ (Page 133)

(i)   The sledge accelerates up the hill.
(ii) The sledge is stationary or moving with constant speed. (Forces in equilibrium.)
(iii) The acceleration is downhill.

Exercise 7A (Page 133)

1 (i)

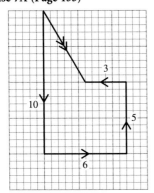

(iii) $3\mathbf{i} - 5\mathbf{j}$; 5.83 N, −59°

2 (i)

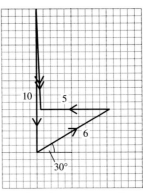

(iii) $0.196\mathbf{i} - 7\mathbf{j}$; 7.00 N, −88.4°

3 (i)

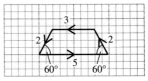

(iii) Equilibrium

4 (i)

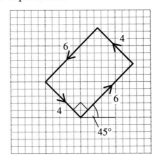

(iii) Equilibrium

5 (i)

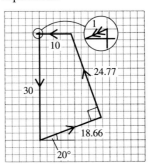

(iii) $-\mathbf{i}$; 1 N down incline

6 (i)

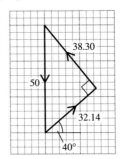

(iii) Equilibrium

? (Page 136)

(i) Not if the string has negligible mass

(ii) No

(iii) (a) $M^2 = M_1^2 + M_2^2$ (b) none

(iv) only if $M < M_1 + M_2$ and
$M_2^2 < M^2 + M_1^2$ (for $M_2 > M_1$)

(v) (a) $\alpha = 120°$ (b) $\alpha > 120°$ (c) $\alpha < 120°$

? (Page 137)

Draw a vertical line to represent the weight,
$10g\,N = 98\,N$. Then add the line of the force T_2 at $45°$
to the horizontal (note the length of this vector is
unknown), and then the line of the force T_1 at $30°$ to
the horizontal ($60°$ to the vertical). C is the point at
which these lines meet.

? (Page 138)

The angles in the triangle are $180° - \alpha$ etc. The sine
rule holds and $\sin(180° - \alpha) = \sin\alpha$ etc.

Exercise 7B (Page 140)

1 (i) 30 N, 36.9°; 65 N, 67.4°

(ii)

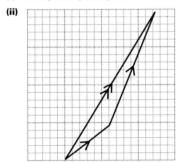

(iii) $49\mathbf{i} + 78\mathbf{j}$; 92.1 N, 57.9°

2 (i)

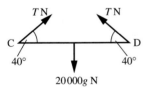

$20000g\,N$

(ii) $T\cos 40°$, $T\sin 40°$; $-T\cos 40°$, $T\sin 40°$

(iv)

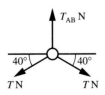

(v) 196 000 N.

(vi) Resolve vertically for the whole system.

3 (i) (a)

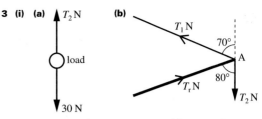

load
30 N

(b)

(ii) Rod: 56.4 N, compression, Cable 1: 59.1 N,
tension

4 (i) 15.04 kg (ii) Both read 10 kg

(iii) Both read 7.64 kg (iv) Method A or C

5 (i)

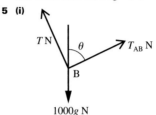

$1000g\,N$

(ii) A force towards the right is required to balance
the horizontal component of T.

(iii)

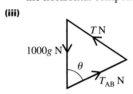

(iv) (a) 9800 N, 13 859 N, 9800 N

(b) 9800 N, 9800 N, 9800 N

6 (ii) (a) $56.1\mathbf{i} + 61.2\mathbf{j}$ (b) 83 N, 47.5°

(iii) 30.8 N, −121°

7 (i) Cable 1: (5638, 2052); Cable 2: ($T_2 \cos 30°$,
$T_2 \sin 30°$)

(ii) 4100 N (iii) 9190 N

8 (i)

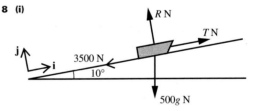

(ii) $-3500\mathbf{i}$, $T\mathbf{i}$, $R\mathbf{j}$, $-851\mathbf{i} - 4826\mathbf{j}$

(iii) 4351 N (iv) 4351 N (v) No

9 (i)

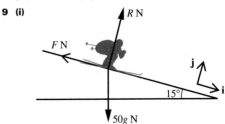

(ii) $-F\mathbf{i}$, $R\mathbf{j}$, $127\mathbf{i} - 473\mathbf{j}$

(iii) 473 N, 127 N (iv) 254 N

10 (i)

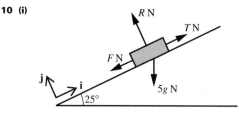

(ii) $T\mathbf{i}, -F\mathbf{i}, R\mathbf{j}, -20.7\mathbf{i} - 44.4\mathbf{j}$

(iii) $T = 29.4, 8.69\,\text{N}$ **(iv)** $1.23\,\text{kg}$

11 (i) $58\mathbf{i} + 15.5\mathbf{j}; 59\mathbf{i} - 10.4\mathbf{j}$

(ii) (a) $117\,\text{N}$ **(b)** $5.11\,\text{N}$

(iii) $97\,\text{N}$ forwards **(iv)** $3\,\text{N}$

12 (i) The wall can only push outwards.

(ii) $T \sin 35° = R \sin \alpha$, $T \cos 35° + R \cos \alpha = 80g$

(iii) $457\,\text{N}$

(iv) $563\,\text{N}$ down along the rope

(v) $563\,\text{N}$

(vi) $1074\,\text{N}$

13 (i) $11.0\,\text{N}, 63.43°$

(ii) A circle with centre A, radius 1 m; No; two parallel forces and a third not parallel cannot form a triangle

14 (i) Horizontal components of tensions must be equal

(ii)

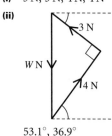

126 N

73.5 N

(iii) $73.5\,\text{N}$

(iv) $T = \dfrac{60 \cos q°}{\cos r°}$; r increases and q decreases so $\cos r°$ decreases and $\cos q°$ increases causing T to increase.

15 (i) $3\,\text{N}, 3\,\text{N}, 4\,\text{N}, 4\,\text{N}$

(ii)

$53.1°, 36.9°$

(iii)

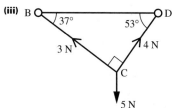

(iv)

The triangle of forces PQR, has sides 3, 4 and W (vertical) so $W < 3 + 4 = 7$

(v)

As W decreases, PQ decreases, but angle $P < 90°$ so $4^2 < W^2 + 3^2$, $W > \sqrt{7}$

❓ (Page 147)

Down the slope.

❓ (Page 149)

Sam and the sledge are a particle. There is no friction and the slope is straight. Friction would reduce both accelerations so Sam would not travel so far on either leg of his journey.

Exercise 7C (Page 151)

1 (i) $1.5\mathbf{i} - \mathbf{j}\,\text{ms}^{-2}$ **(ii)** $1.8\,\text{ms}^{-2}$

2 (i) $4\mathbf{i} + 11\mathbf{j}$ **(ii)** $8\mathbf{i} + 8\mathbf{j}, 2\mathbf{i} + 2\mathbf{j}$

3 (i)

(ii) $11.55\,\text{N}$ **(iii)** $1\,\text{ms}^{-2}$ **(iv)** $0.4\,\text{s}$

4 (i)

(ii) Davies: $-19.3\mathbf{i} + 23\mathbf{j}$; Jones: $30.6\mathbf{i} + 25.7\mathbf{j}$; Total: $11.4\mathbf{i} + 30\mathbf{j}$

(iii) $17\,\text{ms}^{-2}$ at $69°$

(iv) The fish swings sideways as it moves up towards Jones.

5 (i)

(ii) $T\mathbf{i} + 0\mathbf{j}; 0\mathbf{i} + R\mathbf{j}; -30g \sin 30°\mathbf{i} - 30g \cos 30°\mathbf{j}$

(iii) 169.5 N

(iv) The crate slows down to a stop and then starts sliding down the slope.

6 (i)

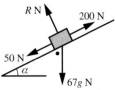

(ii) 13.2°   **(iii)** 0.87 ms^{-2}, 35.9 m   **(iv)** 9.37 ms^{-1}

7 (i)

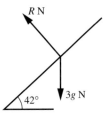

(ii) 6.56 ms^{-2} **(iii)** 1.75 s   **(iv)** 13 N

8 (i)

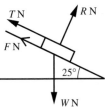

(iv) 2.15 ms^{-2}, 1.7 s

9 (i) The horizontal component of tension in the rope needs a balancing force.

(ii)

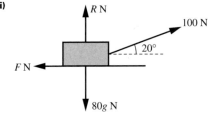

(iii) 94.0 N, 750 N   **(iv)** 128 N   **(v)** 0.144 ms^{-2}

10 (i) 98.5 N, 17.4 N

(ii) The reaction of the rails.

(iii) 98.5 N

(iv) All the forces are constant. 109.3 N, 11.25 m

(v) $ma = T\cos\theta - (100 + 44\sin\theta)$, each term decreases as θ increases, so a decreases.

11 (i)

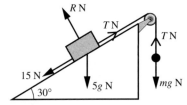

(ii) 4.03 kg, 39.5 N

(iii) 1.75 ms^{-2}, 48.3 N

12 (i) 2.4 ms^{-2}

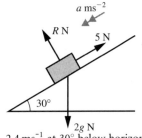

(ii) 2.4 ms^{-1} at 30° below horizontal

(iii) 0.67 s   **(iv)** 1.39 m

13 (i) 6.3 ms^{-2}, No   **(ii)** 6.89 ms^{-1} **(iii)** 1.39 m

Chapter 8

Exercise 8A (Page 159)

1 (i) **(a)** $v = 2 - 2t$ **(b)** 10, 2   **(c)** 1, 11

   (ii) **(a)** $v = 2t - 4$ **(b)** 0, –4   **(c)** 2, –4

   (iii) **(a)** $3t^2 - 10t$ **(b)** 4, 0   **(c)** 0, 4 and $3\frac{1}{3}$, –14.5

2 (i) **(a)** 4   **(b)** 3, 4   **(ii)** **(a)** $12t - 2$ **(b)** 1, –2

   (iii) **(a)** 7   **(b)** –5, 7

3 $v = 4 + t$, $a = 4$

4 (i) **(a)** $v = 15 - 10t$, $a = -10$

   (b)

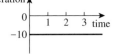

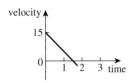

   (c) The acceleration is the gradient of the velocity–time graph.

   (d) The acceleration is constant; the velocity decreases at a constant rate.

   (ii) **(a)** $v = 18t^2 - 36t - 6$, $a = 36t - 36$

   (b)

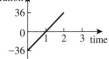

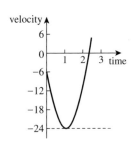

(c) The acceleration is the gradient of the velocity–time graph; velocity is at a minimum when the acceleration is 0.

(d) It starts in the negative direction. v is initially -6 and decreases to -24 before increasing rapidly to zero, where the object turns to move in the positive direction.

Exercise 8B (Page 164)

1 (i) $2t^2 + 3t$

  (ii) $1.5t^4 - \frac{2}{3}t^3 + t + 1$

  (iii) $\frac{7}{3}t^3 - 5t + 2$

2 (i)
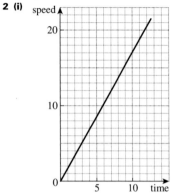

  (ii) $85\,\text{m}$

3 (i) When $t = 6$

  (ii) $972\,\text{m}$

4 (i) $4.47\,\text{s}$

  (ii) $119\,\text{m}$

5 (i) $v = 10t + \frac{3}{2}t^2 - \frac{1}{3}t^3$, $x = 5t^2 + \frac{1}{2}t^3 - \frac{1}{12}t^4$

  (ii) $v = 2 + 2t^2 - \frac{2}{3}t^3$, $x = 1 + 2t + \frac{2}{3}t^3 - \frac{1}{6}t^4$

  (iii) $v = -12 + 10t - 3t^2$, $x = 8 - 12t + 5t^2 - t^3$

? **(Page 164)**

(i); $s = ut + \frac{1}{2}at^2$; $v = u + at$; $a = 4$, $u = 3$.
In the other cases the acceleration is not constant.

? **(Page 165)**

Substituting $at = v - u$ in ② gives
$s = ut + \frac{1}{2}(v - u)\,t + s_0 \Rightarrow s = \frac{1}{2}(u + v)\,t + s_0$ ③;
$v - u = at$ and $v + u = 2(s - s_0)$
$\Rightarrow (v - u)(v + u) = at \times \dfrac{2}{t}(s - s_0)$
$\Rightarrow v^2 - u^2 = 2a(s - s_0)$ ④;
Substituting $u = v - at$ in ② gives $s = vt - \frac{1}{2}at^2 + s_0$

Exercise 8C (Page 165)

1 (i) $15 - 10t$

  (ii) $11.5\,\text{m}, +5\,\text{ms}^{-1}, 5\,\text{ms}^{-1}$; $11.5\,\text{m}, -5\,\text{ms}^{-1}, 5\,\text{ms}^{-1}$

(iii)
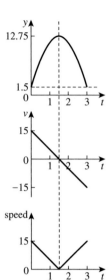

(iv) $3\,\text{s}$

(v) The expression does not equal the distance travelled because of changes in direction. The expression gives the displacement from the origin which equals 0.

2 (i) $-3\,\text{m}, -1\,\text{ms}^{-1}, 1\,\text{ms}^{-1}$

  (ii) (a) $1\,\text{s}$ **(b)** $2.15\,\text{s}$

(iii)
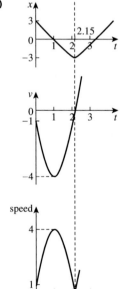

(iv) The object moves in a negative direction from $3\,\text{m}$ to $-3\,\text{m}$ then moves in a positive direction with increasing speed.

3 $2\,\text{s}$

4 (i) $v = 4 + 4t - t^2$, $s = 4t + 2t^2 - \frac{1}{3}t^3$

(ii)

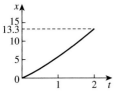

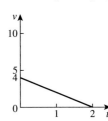

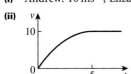

(iii) The object starts at origin and moves in a positive direction with increasing speed reaching a maximum speed of 8 ms⁻¹ after 2 s.

5 (i) 0, 10.5, 18, 22.5, 24

(ii) The ball reaches the hole at 4 s.

(iii) $-3t + 12$ (ms⁻¹)   **(iv)** 0 ms⁻¹   **(v)** -3 ms⁻²

6 (i) Andrew: 10 ms⁻¹, Elizabeth: 9.6 ms⁻¹

(ii)

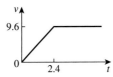

(iii) 11.52 m

(iv) 11.62 s

(v) Elizabeth by 0.05 s and 0.5 m

(vi) Andrew wins

7 (i)

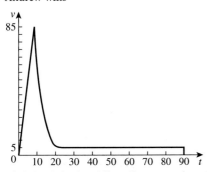

Christine is in free fall until $t = 10$ s then the parachute opens and she slows down to terminal velocity of 5 ms⁻¹.

(ii) 1092 m

(iii) 8.5 ms⁻², $1.6t - 32$, 0 ms⁻², 16 ms⁻²

8 (i) Between P and Q the train speeds up with gradually decreasing acceleration. Between Q and R it is travelling at a constant speed. Between S and T the train is slowing down with a constant deceleration.

(ii) −0.000025, 0.05

(iii) 50 ms⁻¹

(iv) 0 ms⁻¹

(v) $111\frac{2}{3}$ km

9 (i)

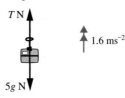

(iii) 855 N

(iv) max tension = 64 N, string breaks

10 (i) $3t^2 - 3$ ms⁻¹, $6t$ ms⁻²

(ii) $t = 1$

(iii)

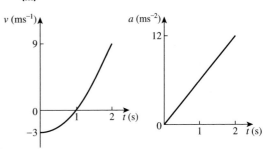

(iv) The object moves towards A and slows down to zero velocity when $t = 1$, then accelerates back towards B reaching a speed of 9 ms⁻¹ when $t = 2$.

(v) 6 m

11 (i) 40

(ii) $s = 0$ when $t = 0$ and 10

(iii) $25 - 5t$

(iv) 62.5 m

(v) In Michelle's model the velocity starts at 25 ms⁻¹ and then decreases. The teacher's model is better because the velocity starts at zero and ends at zero.

12 (i) **(a)** 112 cm   **(b)** 68 cm   **(ii)** $4t$, 16

(iii) $2t^2$, $16t - 32$ **(iv)** $111\frac{1}{9}$ cm, $\frac{8}{9}$ cm less

? (Page 169)

$y = 0$, when $t = 2.4$ and then the projectile hits the ground.

Exercise 8D (Page 175)

1 (i) $4t\mathbf{i} + 8\mathbf{j}$ **(ii)** $(0, 0), (2, 8), (18, 24), (32, 32)$

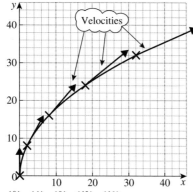

(iii) $\binom{0}{8}, \binom{4}{8}, \binom{8}{8}, \binom{12}{8}, \binom{16}{8}$

(iv) $21.5\,\text{ms}^{-1}$

2 $\mathbf{v} = -4\mathbf{i} - 5\mathbf{j}, \mathbf{a} = 0\mathbf{i} + 0\mathbf{j}$

3 $4.47\,\text{ms}^{-2}, -153°$

4 (i) $\frac{1}{20}t^2\mathbf{i} + \frac{1}{30}t^3\mathbf{j}$

(ii) $5\mathbf{i} + 33.3\mathbf{j}\,\text{m}$

5 $\mathbf{v} = 2t^2\mathbf{i} + (6t - t^2)\mathbf{j}, \mathbf{r} = \frac{2}{3}t^3\mathbf{i} + (3t^2 - \frac{1}{3}t^2)\mathbf{j}$

6 $8.11\,\text{ms}^{-1}$

7 (i) Initial velocity $= 3.54\mathbf{i} - 3.54\mathbf{j}$

(ii) $\mathbf{v} = 8.54\mathbf{i} + 11.46\mathbf{j}; \mathbf{r} = 52.1\mathbf{i} + 14.6\mathbf{j}\,\text{m}$

8 (i) $\mathbf{v} = 15\mathbf{i} + (16 - 10t)\mathbf{j}; \mathbf{a} = -10\mathbf{j}$

(ii) $1.6\,\text{s}$

(iii) $22.8\,\text{ms}^{-1}$

(iv) $y = 2 + \frac{16}{15}x - \frac{1}{45}x^2$

9 (i) $(0, 2), (4, 2), (8, 1.6), (12, 0.8), (16, -0.4)$

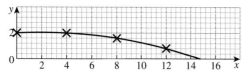

(ii) $20\mathbf{i} + (1 - 10t)\mathbf{j}; 20\mathbf{i} - \mathbf{j}$

(iii) $-10\mathbf{j}$

(iv) $y = 2 + \frac{x}{20} - \frac{1}{80}x^2$

10 (i)

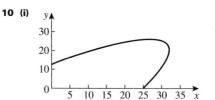

(ii) $5\,\text{s}$

(iii) $33.5\,\text{ms}^{-1}, -117°$

11 (i) A: $v\sin 35°\,\mathbf{i} + v\cos 35°\,\mathbf{j}$; B: $-8.66\mathbf{i} + 5\mathbf{j}$

(ii) A: $vt\sin 35°\,\mathbf{i} + vt\cos 35°\,\mathbf{j}$;
B: $(-8.66t + 5)\,\mathbf{i} + 5t\,\mathbf{j}$

(iii) $6.1\,\text{km h}^{-1}$

(iv) 24.7 minutes

12 (i) $2t^2\mathbf{i} + 4t\mathbf{j}$

(ii) $\frac{2}{3}t^3\mathbf{i} + 2t^2\mathbf{j}$

(iii) $\frac{1}{6}t^4\mathbf{i} + \frac{2}{3}t^3\mathbf{j}$

(iv) $\mathbf{F} = 4\mathbf{i} + 4\mathbf{j}, \mathbf{a} = 8\mathbf{i} + 8\mathbf{j}, \mathbf{v} = \frac{16}{3}\mathbf{i} + 8\mathbf{j}$,
$\mathbf{r} = \frac{8}{3}\mathbf{i} + \frac{16}{3}\mathbf{j}$, speed $9.6\,\text{ms}^{-1}$, $6\,\text{m}$ from O

13 (i)

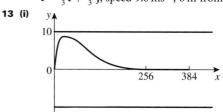

No, maximum $y = 9.48$

(ii) $\mathbf{v} = 8t\mathbf{i} + \frac{1}{8}(64 - 32t + 3t^2)\mathbf{j}$;
$\mathbf{a} = 8\mathbf{i} + \frac{1}{8}(6t - 32)\mathbf{j}, 0 \leqslant t \leqslant 8$;
$\mathbf{v} = 64\mathbf{i}; \mathbf{a} = 0\mathbf{i} + 0\mathbf{j}, 8 < t \leqslant 20$

(iii) $1200\mathbf{i} + (112.5t - 600)\mathbf{j}\,\text{N}$

(iv) air resistance

(v) The motorcycle hits an obstacle and leaves the ground.

14 (i) $16\,\text{cm}$

(ii) $20\,\text{cm}$

(iii) $0\,\text{cm s}^{-1}, 0\,\text{cm s}^{-1}; 20\,\text{cm s}^{-1}, -16\,\text{cm s}^{-1}$

(v) all zero; all components of velocity being zero is safer when the chip is put on the table.

Index

5 Find the remainder when $3x^3 - 2x^2 + 9x + 15$ is divided by $x + 2$.

Review Heinemann Book P3 pages 12–18

6 Prove that $f(x)$, where

$$f(x) \equiv 2x^3 + 5x^2 - 19x - 42$$

is divisible by $2x + 7$.
Hence find the values of x for which $f(x) = 0$.

Review Heinemann Book P3 pages 12–18

7 Expand in ascending powers of x as far as, and including, the term in x^3:

(a) $(1 + 4x)^{-2}$
(b) $(1 - 4x)^{\frac{1}{4}}$

Simplify each coefficient.

Review Heinemann Book P3 pages 20–23

8 Expand $(8 + 3x)^{-\frac{1}{3}}$ in ascending powers of x up to and including the term in x^2 and simplify each coefficient. State the set of values of x for which your series is valid.

Review Heinemann Book P3 pages 20–23

9 (a) Write down the first four terms in the binomial expansion of $(1 + 3x)^n$ in ascending powers of x.
(b) By choosing suitable values of n and x, use the series to find the cube root of 1003 correct to 9 significant figures.

Review Heinemann Book P3 pages 20–23

10 $f(x) \equiv 2x^3 - 3x^2 + Ax + B$, where A and B are constants. When divided by $2x - 7$, $f(x)$ leaves a remainder of -15 and $f(\frac{3}{2}) = 0$.
Factorise $f(x)$ completely.

Review Heinemann Book P3 pages 12–18

Test yourself answers

1 $\dfrac{1}{2x-1} + \dfrac{3}{x+1}$ **2** $\dfrac{1}{x} - \dfrac{3}{x^2+1}$ **3** $\dfrac{1}{x-2} + \dfrac{1}{x^2} + \dfrac{2}{x}$ **4** $A = \frac{1}{3}, B = -\frac{4}{3}, C = -\frac{4}{3}$ **5** -35 **6** $3, -2, -\frac{7}{2}$

7 (a) $1 - 8x + 48x^2 - 256x^3$ (b) $1 - x - \frac{3}{2}x^2 - \frac{7}{2}x^3$ **8** $\frac{1}{2} - \frac{1}{16}x + \frac{1}{64}x^2; |x| < \frac{8}{3}$

9 (a) $1 + 3nx + \dfrac{9}{2}n(n-1)x^2 + \dfrac{9}{2}n(n-1)(n-2)x^3$ (b) $x = 10^{-3}, n = \frac{1}{3} \Rightarrow (1003)^{\frac{1}{3}} = 10.009\,990\,017$ **10** $(x-4)(x+4)(2x-3)$

Differentiation

2

Key points to remember

1 If y is a function of t and t is a function of x, then

$$\frac{dy}{dx} = \frac{dy}{dt} \cdot \frac{dt}{dx}$$

This is called the chain rule.

2 If $y = uv$, where u and v are functions of x, then

$$\frac{dy}{dx} = v\frac{du}{dx} + u\frac{dv}{dx}$$

This is called the product rule.

3 If $y = \dfrac{u}{v}$, where u and v are functions of x, then

$$\frac{dy}{dx} = \frac{v\dfrac{du}{dx} - u\dfrac{dv}{dx}}{v^2}$$

This is called the quotient rule.

4 The chain rule is used to find connected rates of change; for example:

$$\frac{dV}{dt} = \frac{dV}{dr} \cdot \frac{dr}{dt}$$

5 You should know that $\lim\limits_{x \to 0}\left(\dfrac{\sin x}{x}\right) = 1$.

6 $\dfrac{dy}{dx} = \dfrac{1}{\dfrac{dx}{dy}}$

7 For functions given implicitly, remember that

$$\frac{d}{dx}(y^2) = 2y\frac{dy}{dx}; \frac{d}{dx}(xy) = y + x\frac{dy}{dx},$$

obtained from the chain rule and product rule respectively.

8 For a curve given parametrically by the equations

$$x = f(t), \quad y = g(t)$$

$$\frac{dx}{dt} = f'(t), \quad \frac{dy}{dt} = g'(t)$$

$$\frac{dy}{dx} = \frac{dy}{dt} \cdot \frac{dt}{dx} = \frac{\dfrac{dy}{dt}}{\dfrac{dx}{dt}} = \frac{g'(t)}{f'(t)}$$

9 Memorise these standard formulae for derivatives:
(a, b, k, n are constants)

y or $f(x)$	$\dfrac{dy}{dx}$ or $f'(x)$
x^n	nx^{n-1}
$(ax + b)^n$	$na(ax + b)^{n-1}$
e^x	e^x
e^{ax+b}	ae^{ax+b}
a^x	$a^x(\ln a)$
$\ln x$	$\dfrac{1}{x}$
$\ln(ax + b)$	$\dfrac{a}{ax + b}$
$\sin x$	$\cos x$
$\cos x$	$-\sin x$
$\tan x$	$\sec^2 x$
$\cot x$	$-\operatorname{cosec}^2 x$
$\sec x$	$\sec x \tan x$
$\operatorname{cosec} x$	$-\operatorname{cosec} x \cot x$
$\sin kx$	$k \cos kx$
$\sin^n x$	$n \sin^{n-1} x \cos x$
$\sin^n kx$	$nk \sin^{n-1} kx \cos kx$
$u \pm v$	$\dfrac{du}{dx} \pm \dfrac{dv}{dx}$
uv	$v\dfrac{du}{dx} + u\dfrac{dv}{dx}$
$\dfrac{u}{v}$	$\dfrac{v\dfrac{du}{dx} - u\dfrac{dv}{dx}}{v^2}$
y^n	$ny^{n-1}\dfrac{dy}{dx}$

Example 1

Find $\dfrac{dy}{dx}$ when $y =$

(a) e^{x^3},   **(b)** $x^{\frac{3}{2}}\ln x$,   **(c)** $\dfrac{x^3+2}{x^2+5}$

Answer

(a) $y = e^{x^3}$, put $x^3 = t$

Then $y = e^t$, $t = x^3$

So $\dfrac{dy}{dt} = e^t$, $\dfrac{dt}{dx} = 3x^2$ | Using **9** |

$\dfrac{dy}{dx} = \dfrac{dy}{dt} \cdot \dfrac{dt}{dx} = e^t . 3x^2$ | Using **1** |

But $t = x^3 \Rightarrow \dfrac{dy}{dx} = 3x^2 e^{x^3}$ | Always ensure the final answer is in terms of x. |

(b) $y = x^{\frac{3}{2}}\ln x \equiv uv$ | Recognise the product. |

$u = x^{\frac{3}{2}},$ $v = \ln x$

$\dfrac{du}{dx} = \tfrac{3}{2}x^{\frac{1}{2}},\ \dfrac{dv}{dx} = \dfrac{1}{x}$ | Using **9** |

$\dfrac{dy}{dx} = \tfrac{3}{2}x^{\frac{1}{2}}\ln x + x^{\frac{3}{2}}\left(\dfrac{1}{x}\right) = \tfrac{3}{2}x^{\frac{1}{2}}\ln x + x^{\frac{1}{2}}$ | Using **2** |

(c) $y = \dfrac{x^3+2}{x^2+5} \equiv \dfrac{u}{v}$ | Recognise the quotient. |

$u = x^3 + 2,$ $v = x^2 + 5$

$\dfrac{du}{dx} = 3x^2,$ $\dfrac{dv}{dx} = 2x$ | Using **9** |

$\dfrac{dy}{dx} = \dfrac{3x^2(x^2+5) - 2x(x^3+2)}{(x^2+5)^2}$ | Using **3** |

$= \dfrac{3x^4 + 15x^2 - 2x^4 - 4x}{(x^2+5)^2} = \dfrac{x^4 + 15x^2 - 4x}{(x^2+5)^2}$

Example 2

Find an equation of the tangent at the point where $x = \frac{\pi}{6}$ on the curve with equation $y = \cos^3 2x$.

Answer

At $x = \frac{\pi}{6}$, $y = \cos^3\frac{\pi}{3} = \left(\frac{1}{2}\right)^3 = \frac{1}{8}$

$y = \cos^3 2x,\ \dfrac{dy}{dx} = 3\cos^2 2x \dfrac{d}{dx}(\cos 2x)$ | Using **1** and **9** |

$= 3\cos^2 2x(-2\sin 2x)$ | Using **1** again |

At $x = \frac{\pi}{6}$, $\dfrac{dy}{dx} = -6\cos^2\frac{\pi}{3}\sin\frac{\pi}{3} = -\dfrac{3\sqrt{3}}{4}$

> $\cos\frac{\pi}{3} = \frac{1}{2},\ \sin\frac{\pi}{3} = \dfrac{\sqrt{3}}{2}$

Equation of the tangent is $y - \frac{1}{8} = -\dfrac{3\sqrt{3}}{4}\left(x - \frac{\pi}{6}\right)$

> See Book P1 page 144.

Example 3

Given that $y = e^{-2x}\sec 3x$, prove that $\dfrac{dy}{dx} = y(3\tan 3x - 2)$.

Answer

Multiply the given equation by e^{2x} to obtain

$$e^{2x}y = \sec 3x,\quad e^{2x}y \equiv uv$$

$$u = e^{2x},\qquad v = y$$

$$\frac{du}{dx} = 2e^{2x},\ \frac{dv}{dx} = \frac{dy}{dx}$$

> Using **1** and **9**

So: $\dfrac{d}{dx}(e^{2x}y) = 2ye^{2x} + e^{2x}\dfrac{dy}{dx}$

> Using **2**

Also: $\dfrac{d}{dx}(\sec 3x) = 3\sec 3x\tan 3x$

> Using **1** and **9**

We now have $\dfrac{d}{dx}(e^{2x}y) = \dfrac{d}{dx}(\sec 3x)$

$$\Rightarrow 2ye^{2x} + e^{2x}\frac{dy}{dx} = 3\sec 3x\tan 3x$$

> Using **7**

But $e^{2x}y = \sec 3x$

$$\Rightarrow 2ye^{2x} + e^{2x}\frac{dy}{dx} = 3ye^{2x}\tan 3x$$

Divide by $e^{2x} \Rightarrow \dfrac{dy}{dx} = 3y\tan 3x - 2y = y(3\tan 3x - 2)$

Example 4

The variables x and y are related by the equation $y^2 = x^5$ and x is increasing at a rate of 0.1 units per second. Find the rate at which y is increasing when $x = 4$.

Answer

Differentiating $y^2 = x^5$ with respect to time t

$$2y\frac{dy}{dt} = 5x^4\frac{dx}{dt}$$

> Using **1**, **4** and **7**

When $x = 4$, $y = 32$ and $\dfrac{dx}{dt} = 0.1$

$$\frac{dy}{dt} = \frac{5 \times 256 \times 0.1}{2 \times 32} = 2 \text{ units per second}$$

Rate of increase of y when $x = 4$ is 2 units per second.

Example 5

A curve is given by the equations

$$x = 2\sin t,\ y = \cos 2t,\ 0 \leqslant t < \tfrac{1}{2}\pi$$

where t is a parameter.

Find $\dfrac{\mathrm{d}y}{\mathrm{d}x}$ in terms of t, simplifying your answer.

Answer

$$\frac{\mathrm{d}x}{\mathrm{d}t} = 2\cos t$$

Using ▉9▉

$$\frac{\mathrm{d}y}{\mathrm{d}t} = -2\sin 2t$$

Using ▉1▉ and ▉9▉

$$\frac{\mathrm{d}y}{\mathrm{d}x} = \frac{\dfrac{\mathrm{d}y}{\mathrm{d}t}}{\dfrac{\mathrm{d}x}{\mathrm{d}t}} = \frac{-2\sin 2t}{2\cos t}$$

Using ▉8▉

$$= \frac{-\sin 2t}{\cos t}$$

$$= \frac{-2\sin t\cos t}{\cos t}$$

Using $\sin 2t \equiv 2\sin t\cos t$

$$= -2\sin t$$

Revision exercise 2

Differentiate each of the following:

1 $\cos\frac{5}{2}x$ **2** $-\tan^2 2x$ **3** $(3x - 1)^{\frac{1}{3}}$

4 $\ln(4x + 3)$ **5** $x^2\sqrt{(4x + 1)}$ **6** $e^{\frac{1}{2}x}\sin 3x$

7 $x\ln(\sec x + \tan x)$ **8** $\dfrac{x^2 - 3}{x^3 + 2}$

9 $\dfrac{e^{x^2}}{\tan 3x}$ **10** $\dfrac{1 - \cos 2x}{1 + \cos 2x}$

11 $\ln(\operatorname{cosec} x - \cot x)$ **12** $\ln\!\left(\dfrac{4 - x^2}{4 + x^2}\right)$

In questions 13–16, find $\dfrac{\mathrm{d}y}{\mathrm{d}x}$ in terms of t when:

13 $x = 4\cos t,\ y = 7\sin t$

14 $x = e^t\tan t,\ y = e^t\sec t$

15 $x = 9t^3,\ y = 18t^2$

16 $x = 2t - \cos 2t,\ y = 2\sin t + \sin 2t$

In questions 17–20, find $\dfrac{dy}{dx}$ in terms of x and y and find an equation of the tangent to the curve with equation $f(x, y) = 0$ at the point given.

17 $f(x, y) \equiv x^3 + y^3 - 35$ at $(2, 3)$

18 $f(x, y) \equiv 3x^2 + 2xy - y^2$ at $(1, 3)$

19 $f(x, y) \equiv x^2 + y^2 - 14x - 2y + 40$ at $(4, 2)$

20 $f(x, y) \equiv y^2 - x(5 - x)^2$ at $(1, -2)$

21 Find an equation of the normal at $(-1, \frac{1}{2})$ to the curve with equation $y = \dfrac{x^2}{x^2 + 1}$.

22 The normal to the curve with equation $y^2 = 4x$ at $P(9, 6)$ meets the curve again at Q. Find the coordinates of Q.

23 For the curve C with equations
$$x = \sin t, \quad y = 2\cos 2t, \quad 0 \leqslant t < \tfrac{\pi}{2}$$
where t is a parameter, find an equation of the tangent to C at the point where $t = \tfrac{\pi}{3}$.

24 Find the coordinates of each turning point on the curve with equation $y = \dfrac{x + 2}{(x + 1)(2 - x)}$.

25 Given that $y = 4x(x + 1)^{\frac{1}{2}}$ and that x is changing at the constant rate of $0.2\,\text{cm}\,\text{s}^{-1}$, find the rate of change of y when $x = 3$.

26 By using differentiation find approximate values for

 (a) $(27.005)^{\frac{2}{3}}$ **(b)** $\dfrac{1}{(5.01)^3}$ **(c)** $(255.8)^{\frac{1}{4}}$

Test yourself	What to review
	If your answer is incorrect:
1 Given that $y = \tan^3 2x$, find the value of $\dfrac{dy}{dx}$ at $x = \tfrac{\pi}{8}$.	*Review Heinemann Book P3 pages 28–30 and 44–49*

2 Find an equation of the normal to the curve with equation $50y = x^2\sqrt{(2x-1)}$ at the point where $x = 5$.

Review Heinemann Book P1 page 144 and Heinemann Book P3 pages 31–32

---

3 Differentiate $\dfrac{x^3}{1 + e^{-2x}}$.

Review Heinemann Book P3 pages 28–30 and 33–34

---

4 Given that $y = x \ln x$, $x > 0$, find the minimum value of y.

Review Heinemann Book P1 pages 140–143 and Heinemann Book P3 pages 31–32

---

5 A curve has parametric equations

$$x = 3t^3, \quad y = t^2 - 2,$$

where t is a parameter.

Find the value of $\dfrac{dy}{dx}$ at the point where $t = -1$.

Review Heinemann Book P3 pages 54–56

---

6 Given that $y^2 = A \sin 3x$, form a first order differential equation that does not contain the constant A.

Review Heinemann Book P3 pages 51–54 and 44–49

---

7 A variable rectangle is inscribed in a circle of diameter 20 cm. At a certain instant one side is 16 cm long and it is increasing at a constant rate of 0.75 cm s^{-1}.
Find the rate at which the shorter sides are changing.

Review Heinemann Book P3 pages 35–37

---

8 Given that $ky = 4^x$, find a first order differential equation which does not contain the constant k.

Review Heinemann Book P3 pages 51–53

Test yourself answers

1 12   **2** $y - \frac{3}{2} = -\frac{30}{23}(x - 5)$ **3** $\dfrac{3x^2(1 + e^{-2x}) + 2x^3 e^{-2x}}{(1 + e^{-2x})^2}$ **4** Min $y = -\dfrac{1}{e}$ (occurs at $x = e^{-1}$)   **5** $-\frac{2}{9}$ **6** $3y^2 \cos 3x - 2y\dfrac{dy}{dx}\sin 3x = 0$

7 -1 cm s^{-1} (i.e. decreasing at a rate of 1 cm s^{-1})   **8** $\dfrac{dy}{dx} - y \ln 4 = 0$

Coordinate geometry in the *xy*-plane

3

Key points to remember

1 The circle, centre the origin, radius a, has equation

$$x^2 + y^2 = a^2$$

2 The circle, centre (α, β), radius r has equation

$$(x - \alpha)^2 + (y - \beta)^2 = r^2$$

3 $x^2 + y^2 + 2gx + 2fy + c = 0$ is the equation of a circle, centre $(-g, -f)$ with radius $\sqrt{(g^2 + f^2 - c)}$.

4 The tangent at (h, k) to the circle with equation $x^2 + y^2 = a^2$ has equation

$$hx + ky = a^2$$

5 The tangent at (h, k) to the circle with equation $x^2 + y^2 + 2gx + 2fy + c = 0$ has equation

$$hx + ky + g(h + x) + f(k + y) + c = 0$$

6 Before attempting to sketch a curve given by a cartesian equation you should consider the following:
   (a) Where does the curve cut the axes?
   (b) Where are the asymptotes, if any?
   (c) What happens as $x \to \pm \infty$?
   (d) Has the curve any symmetry?
   (e) Are there any points at which the curve is undefined?
   (f) Where do the stationary points occur?

7 If you are asked to sketch a curve given by parametric equations, try to eliminate the parameter between $x = f(t)$ and $y = g(t)$ to obtain the cartesian equation of the curve and then proceed as in **6**.

---

8 If you are asked to sketch a curve given by parametric equations and you either cannot obtain the cartesian equation or the cartesian equation is more complicated than the parametric equations, then sketch the curve from the parametric equations, asking the following questions:
(a) Where does the curve cut the axes?
(b) Does the curve have any symmetry?
(c) Are there any points at which the curve is undefined?
If all else fails, plot a few points.

9 For the circle with centre the origin and radius a, suitable parametric equations are

$$x = a\cos t, \qquad y = a\sin t, \qquad 0 \leqslant t < 2\pi.$$

10 For the circle with centre (α, β) and radius r, suitable parametric equations are

$$x = \alpha + r\cos\theta, \qquad y = \beta + r\sin\theta, \qquad 0 \leqslant \theta < 2\pi.$$

Example 1

For the circle with equation $x^2 + y^2 - 8x + 12y + 3 = 0$ find the coordinates of the centre and the radius.

Answer

The equation of the circle can be rewritten as

$$x^2 - 8x + 16 + y^2 + 12y + 36 = 49$$

by completing the square on $x^2 - 8x$ and on $y^2 + 12y$.
That is $\qquad (x - 4)^2 + (y + 6)^2 = 7^2$
and from this equation you can immediately give the centre of the circle as $(4, -6)$ and the radius as 7.

Using **2**

Or $\qquad 2g = -8, 2f = 12, g^2 + f^2 - c = (-4)^2 + 6^2 - 3 = 49$
so centre is $(4, -6)$ and radius $= 7$.

Using **3**

Example 2

A circle is drawn with $A(x_1, y_1)$ and $B(x_2, y_2)$ at the ends of a diameter. Find an equation of this circle.

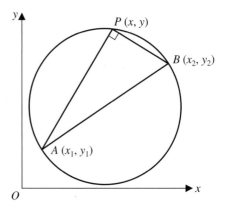

Answer

Take any point $P(x, y)$ on the circle, as shown.
From GCSE work you know that $\angle APB = 90°$, the angle in a semicircle.

Gradient of $AP = \dfrac{y - y_1}{x - x_1}$

Gradient of $BP = \dfrac{y - y_2}{x - x_2}$

Since $\angle APB = 90°$, $\dfrac{(y - y_1)(y - y_2)}{(x - x_1)(x - x_2)} = -1$

> For perpendicular lines of gradients m_1 and m_2, $m_1 m_2 = -1$.

That is $(y - y_1)(y - y_2) = -(x - x_1)(x - x_2)$ and you can take the equation of the circle with AB as diameter to be

$$(x - x_1)(x - x_2) + (y - y_1)(y - y_2) = 0.$$

Example 3

Show that the points $A(2, 2)$, $B(7, -3)$ and $C(-2, 0)$ are each at the same distance from the point $D(2, -3)$.
Hence find an equation of the circle passing through the points A, B and C.

Answer

Using the formula $d^2 = (x_1 - x_2)^2 + (y_1 - y_2)^2$ for the distance d between (x_1, y_1) and (x_2, y_2), you have

$$DA^2 = (2 - 2)^2 + (-3 - 2)^2 = 25$$
$$DB^2 = (2 - 7)^2 + (-3 - -3)^2 = 25$$
$$DC^2 = (2 - -2)^2 + (-3 - 0)^2 = 16 + 9 = 25$$

Using Pythagoras' theorem

which confirms that $DA = DB = DC$ and that D is the centre of the circle through A, B and C.
An equation of the circle is

$$(x - 2)^2 + (y + 3)^2 = 25$$

Using **2**

or

$$x^2 + y^2 - 4x + 6y - 12 = 0.$$

Example 4

Sketch the curve given by the equations

$$x = t + 1, \quad y = 4 - t^2,$$

where t is a parameter.
Write on your sketch the coordinates of the turning point and where the curve meets the axes.

Answer

Note that $t = x - 1$ and since $y = 4 - t^2$

$$y = 4 - (x - 1)^2 = -x^2 + 2x + 3$$

Using **7**

So the cartesian equation is easily obtained and can be used to sketch the quadratic curve.
When $\quad x = 0, \quad y = 3$
and when $\quad y = 0, \quad -x^2 + 2x + 3 = 0 \Rightarrow x = 3$ or $x = -1$

Using **6**

The curve is a parabola as shown with maximum point $(1, 4)$ and meeting the axes at $(0, 3)$, $(-1, 0)$ and $(3, 0)$.

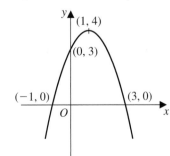

The turning point is quickly found using the symmetry of the parabola.

Example 5

Find an equation of the tangent at $(3, -4)$ on the circle with equation $x^2 + y^2 = 25$.

Answer

Differentiating with respect to x

$$2x + 2y\frac{dy}{dx} = 0$$

> Using implicit differentiation.

At $(3, -4)$: $\quad 6 - 8\dfrac{dy}{dx} = 0 \Rightarrow \dfrac{dy}{dx} = \frac{3}{4}$

The tangent has gradient $\frac{3}{4}$ and passes through $(3, -4)$.

An equation of the tangent is $y + 4 = \frac{3}{4}(x - 3)$.

Or: tangent to circle at $(3, -4)$ has equation $3x - 4y = 25$.

> Using **2**

Revision exercise 3

In questions 1–4 find equations for the circle
(a) in cartesian form,   **(b)** in a parametric form.

1 Centre $(0, 0)$, radius 6.

2 Centre $(2, 1)$, radius 6.

3 Centre $(-3, -5)$, radius 4.

4 Through the points $(2, 0)$, $(8, 0)$ and $(5, 9)$.

5 Find, in cartesian form, an equation of the circle which touches the y-axis at the point $(0, 6)$ and also passes through the point $(1, 3)$.

6 Find, in cartesian form, an equation of the circle which has its centre on the x-axis, and passes through both the origin O and the point $(2, -2)$.

7 Find, in cartesian form, an equation of the circle which has the points $(7, -9)$ and $(-1, 1)$ at the ends of a diameter.

In questions 8–10 find the coordinates of the centre and the radius of the circle with equations:

8 $x^2 + y^2 - 7x + 24y = 0$

9 $4x^2 + 4y^2 - 20x + 28y - 73 = 0$

10 $x = 3 - 4\cos t, \quad y = 5 + 4\sin t, \quad 0 \leqslant t < 2\pi$

11 Find the coordinates of the centre and the radius of the circle with equation
$$x^2 + y^2 - 8x - 2y + 8 = 0$$
By calculation determine whether the point $P(6.4, 2.9)$ lies inside or outside the circle.

12 Show that the lines with equations $2x - 11 = 0$ and $2y + 1 = 0$ are tangents to the circle with equation
$$x^2 + y^2 - 4x - 6y + \tfrac{3}{4} = 0.$$

In questions 13–16, find a cartesian equation for the curve whose parametric equations are:

13 $x = \dfrac{4}{t}, y = 7t$

14 $x = 6t, y = 6t^2 + 9$

15 $x = 120t - 16t^2, y = 20t$

16 $x = \sec t, y = \tan t \quad \left(-\tfrac{\pi}{2} < t < \tfrac{\pi}{2}\right)$

In questions 17–20, sketch the curves with equations:

17 $x^2 + 3y^2 = 9$

18 $y = x^4 - x^2$

19 $y = \dfrac{4x}{x^2 + 1}$

20 $y = \dfrac{x^2 + 1}{4x}$

21 Prove that for all values of the parameter t, the line with equation $x - ty + 2(t^2 - 1) = 0$ is a tangent to the curve with equation $y^2 = 8(x - 2)$.

22 Show that $P(2t, t^2)$ lies on the curve with equation $4y - x^2 = 0$ for all values of t.

At the point A on the curve $t = -3$. The tangent to the curve at A meets the x-axis at the point B.

Find an equation of the circle that has AB as a diameter.

23 Prove that the curve with equation $y = xe^{-x}$ has one maximum point. Sketch the curve given that $xe^{-x} \to 0$ as $x \to \infty$.

24 The curve C_1 has equation $2y^2 - x = 0$.

The curve C_2 has parametric equations
$$x = 4t, y = \dfrac{4}{t}$$

Sketch the curves C_1 and C_2 on the same diagram and calculate the coordinates of their points of intersection.

25 Given that a curve has cartesian equation
$$y = \frac{10x + 15}{x^2 + 4}$$
(a) find the set of values of y for which x takes all real values,

(b) sketch the curve.

Test yourself	What to review
	If your answer is incorrect:
1 Find an equation of the circle (a) with radius 5 and passing through $O(0, 0)$ and $B(0, 6)$, (b) with centre at $(2, 7)$ and passing through the point $(-8, 31)$.	*Review Heinemann Book P3 pages 61–66*
2 The circle C_1 has equation $$x^2 + y^2 - 4x + 6y - 3 = 0.$$ Find an equation of the circle C_2 which has the same centre as C_1 and passes through the point $(5, -6)$.	*Review Heinemann Book P3 pages 61–66*
3 Find an equation of the circle having $A(5, -9)$ and $B(-16, 11)$ as the ends of a diameter.	*Review Heinemann Book P3 page 64, Example 3*
4 Find an equation of the circle passing through the points $(0, 5)$, $(0, 6)$ and $(6, 8)$. Find an equation of the tangent at $(6, 8)$ to this circle.	*Review Heinemann Book P3 pages 64–65, Example 4 and pages 66–68*
5 Show that the line with equation $3x - 4y = 10$ is a tangent to the circle with equation $$2x^2 + 2y^2 - 5x = 0.$$ Find the coordinates of the point of contact.	*Review Heinemann Book P3 pages 66–68*
In the curves you sketch for questions 6–8, label the coordinates of any points where a curve meets the axes.	
6 On the same axes sketch the curves with equations (a) $y = x(x + 1)(x - 3)$ (b) $y^2 = x(x + 1)(x - 3)$	*Review Heinemann Book P3 pages 71–80*

7 On the same axes sketch the curves with equations

(a) $y = \dfrac{x - 3}{x + 3}$

(b) $y = \dfrac{x + 3}{x - 3}$

Review Heinemann Book P3 pages 71–80

---

8 Sketch the curve with equations

$$x = \tfrac{1}{4}(t^2 - 9), \ y = t,$$

where t is a parameter.
Find a cartesian equation of this curve.

Review Heinemann Book P3 pages 81–87

---

9 Find a cartesian equation for the curve given parametrically by

$$x = 3 \tan t, \quad y = 2 \sec t, \quad 0 \leqslant t \leqslant \pi$$

Sketch this curve.

Review Heinemann Book P3 pages 81–87

---

10 Show that the line with equation $ax + by + c = 0$ touches the curve with equation $y^2 = 4x$ if $b^2 + ac = 0$.
Hint: Eliminate y to obtain a quadratic equation in x. This equation has equal roots for a tangent.

Review Heinemann Book P3 pages 81–87

Test yourself answers

1 (a) $x^2 + y^2 - 8x - 6y = 0$ **(b)** $(x - 2)^2 + (y - 7)^2 = 676$ **2** $(x - 2)^2 + (y + 3)^2 = 18$ **3** $(x - 5)(x + 9) + (y + 16)(y - 11) = 0$

4 $x^2 + y^2 - 7x - 11y + 30 = 0; x + y = 14$ **5** $(2, 1)$ **8** $y^2 = 4x + 9$ **9** $\dfrac{y^2}{4} - \dfrac{x^2}{9} = 1$

Integration

4

Key points to remember

1 Standard results (to be memorised):

$$\int \sin x \, dx = -\cos x + C$$

$$\int \sin (ax + b) \, dx = -\frac{1}{a}\cos (ax + b) + C$$

$$\int \cos x \, dx = \sin x + C$$

$$\int \cos (ax + b) \, dx = \frac{1}{a}\sin (ax + b) + C$$

$$\int (ax + b)^n dx = \frac{1}{a(n + 1)}(ax + b)^{n+1} + C, \, n \neq -1$$

$$\int \frac{1}{ax + b}\, dx = \frac{1}{a}\ln |ax + b| + C$$

$$\int e^{ax+b}\, dx = \frac{1}{a}e^{ax+b} + C$$

$$\int \tan x \, dx = \ln |\sec x| + C$$

$$\int \cot x \, dx = \ln |\sin x| + C$$

$$\int \ln x \, dx \quad (x > 0) = x \ln x - x + C$$

$$\int \sec^2 x \, dx = \tan x + C$$

$$\int \operatorname{cosec}^2 x \, dx = -\cot x + C$$

$$\int \sec x \tan x \, dx = \sec x + C$$

$$\int \operatorname{cosec} x \cot x \, dx = -\operatorname{cosec} x + C$$

2 $\int f(x)f'(x)dx = \frac{1}{2}[f(x)]^2 + C$

3 $\int [f(x)]^n f'(x)dx = \frac{1}{n+1}[f(x)]^{n+1} + C, \ n \neq -1$

4 $\int \frac{f'(x)}{f(x)}dx = \ln|f(x)| + C$

5 Integration by parts gives
$$\int v\frac{du}{dx}dx = uv - \int u\frac{dv}{dx}dx$$

6 $y = \int f(x)dx + C$ is called the general solution of the differential equation $\frac{dy}{dx} = f(x)$, once the integration has been completed.

7 The differential equation $\frac{dy}{dx} = f(x)\,g(y)$ is called a first order equation with variables separable. Its general solution is
$$\int \frac{1}{g(y)}dy = \int f(x)dx + C$$
provided that $\frac{1}{g(y)}$ can be integrated with respect to y and that $f(x)$ can be integrated with respect to x.

Example 1
Find

(a) $\int \frac{1}{5x-2}dx$,

(b) $\int e^{1-3x}dx$.

Answer

(a) Consider $\frac{d}{dx}[\ln(5x-2)] = \frac{1}{5x-2} \times \frac{d}{dx}(5x-2)$

See Book P3 pages 27–30.

$$= \frac{5}{5x-2}$$

So $\int \frac{1}{5x-2}dx = \frac{1}{5}\ln|5x-2| + C$

Using **1**

(b) Consider $\dfrac{d}{dx}(e^{1-3x}) = e^{1-3x}\dfrac{d}{dx}(1-3x)$

See Book P3 pages 27–30.

$$= -3e^{1-3x}$$

So $\displaystyle\int e^{1-3x}dx = -\tfrac{1}{3}e^{1-3x} + C$

Using **1**

Example 2
Use integration by parts to find $\displaystyle\int \cos^2 x\, dx$.

Answer

$$\int \cos^2 x\, dx = \int \cos x \cos x\, dx \equiv \int u\frac{dv}{dx}\, dx$$

$u = \cos x$	$v = \sin x$
$\dfrac{du}{dx} = -\sin x$	$\dfrac{dv}{dx} = \cos x$

Use a table like this.

$$\int u\frac{dv}{dx}\, dx = uv - \int v\frac{du}{dx}\, dx$$

Using **5**

$$\int \cos^2 x\, dx = \cos x \sin x - \int (-\sin x)\sin x\, dx$$

$$= \cos x \sin x + \int \sin^2 x\, dx$$

$$= \cos x \sin x + \int (1 - \cos^2 x)\, dx$$

Using $\sin^2 x \equiv 1 - \cos^2 x$

So $\displaystyle\int \cos^2 x\, dx = \cos x \sin x + \int 1\, dx - \int \cos^2 x\, dx$

Using
$$\int (s + t)\, dx = \int s\, dx + \int t\, dx$$

$$2\int \cos^2 x\, dx = \cos x \sin x + x$$

$$\int \cos^2 x\, dx = \tfrac{1}{2}[x + \cos x \sin x] + C$$

Example 3

Evaluate $\displaystyle\int_1^4 \frac{x}{1 + \sqrt{x}}\, dx$.

See Book P3 pages 110–113, on integration using substitutions.

Answer

Put $1 + \sqrt{x} = t$, then $\sqrt{x} = t - 1 \Rightarrow x = (t-1)^2$
when $x = 1$, $t = 2$
when $x = 4$, $t = 3$

Note how to change the limits from x to t.

So $\displaystyle\int_1^4 \frac{x}{1 + \sqrt{x}}\, dx = \int_2^3 \frac{(t-1)^2}{t}\frac{dx}{dt}\, dt$

But $\dfrac{dx}{dt} = 2(t-1)$

Using chain rule of differentiation.

So $\displaystyle\int_1^4 \dfrac{x}{1+\sqrt{x}}\,dx = \int_2^3 \dfrac{(t-1)^2 2(t-1)}{t}\,dt$

$\qquad\qquad\qquad\quad = \displaystyle\int_2^3 \dfrac{2(t-1)^3}{t}\,dt$

$\qquad\qquad\qquad\quad = \displaystyle\int_2^3 \dfrac{2}{t}(t^3 - 3t^2 + 3t - 1)\,dt$

Using
$(x-y)^3 \equiv x^3 - 3x^2 y + 3xy^2 - y^3$

$\qquad\qquad\qquad\quad = \displaystyle\int_2^3 (2t^2 - 6t + 6 - 2t^{-1})\,dt$

$\qquad\qquad\qquad\quad = \left[\tfrac{2}{3}t^3 - 3t^2 + 6t - 2\ln t\right]_2^3$

Using **1**

$\qquad\qquad\qquad\quad = \left(\tfrac{2}{3}(3^3) - 3(3)^2 + 18 - 2\ln 3\right) - \left(\tfrac{2}{3}(2^3) - 3(2)^2 + 12 - 2\ln 2\right)$

$\qquad\qquad\qquad\quad = \tfrac{11}{3} + 2\ln\tfrac{2}{3}$

Example 4

Find

(a) $\displaystyle\int \cot x \operatorname{cosec}^2 x\,dx,$

(b) $\displaystyle\int \dfrac{2x+1}{x^2+x}\,dx.$

Answer

(a) Notice that $\dfrac{d}{dx}(\cot x) = -\operatorname{cosec}^2 x$ and so we have a function

$(\cot x)$ and its derivative $(-\operatorname{cosec}^2 x)$.

$\displaystyle\int f(x)\,f'(x)\,dx \equiv -\int \cot x\,(-\operatorname{cosec}^2 x)\,dx = -\tfrac{1}{2}\cot^2 x + C$

Using **2**

(b) Notice that $\dfrac{d}{dx}(x^2 + x) = 2x + 1$ and so we have a function

$(x^2 + x)$ and its derivative $(2x + 1)$.

$\displaystyle\int \dfrac{f'(x)}{f(x)}\,dx \equiv \int \dfrac{2x+1}{x^2+x}\,dx = \ln|x^2 + x| + C$

Using **4**

Example 5

Given that $\dfrac{dy}{dx} = \tan^2 x$ and $y = 1$ at $x = \tfrac{\pi}{4}$, find y in terms of x.

Answer

Since $\sec^2 x \equiv 1 + \tan^2 x \Rightarrow \tan^2 x \equiv \sec^2 x - 1$

$\qquad\qquad \dfrac{dy}{dx} = \sec^2 x - 1$

Using an identity to move to a function which can be integrated.

Integrating,

$$y = \tan x - x + C,$$

which is the general solution of the given equation.

Using **6**

Now $y = 1$ at $x = \frac{\pi}{4}$

$$\Rightarrow 1 = \tan\frac{\pi}{4} - \frac{\pi}{4} + C$$

Using the boundary conditions.

$$\therefore C = \frac{\pi}{4}$$

The specific solution is $y = \tan x - x + \frac{\pi}{4}$.

Worked examination question 1 [E]

An electrically charged body loses its charge, Q coulombs, at a rate kQ coulombs per second, where k is a constant. Write down a differential equation involving Q and t, where t seconds is the time since the discharge started. Solve this equation for Q, given that the initial charge was Q_0 coulombs.

Given that $Q_0 = 0.001$, and $Q = 0.0005$ when $t = 10$, find the value of Q when $t = 20$.

Answer

From the given data $\dfrac{dQ}{dt} = -kQ$ is the differential equation required.

Separating the variables gives $\dfrac{1}{Q}\,dQ = -k\,dt$

Using **7**

Integrating $\qquad \ln Q = -kt + C$

Using **1**

We are given that $Q = Q_0$ at $t = 0$

so $\qquad C = \ln Q_0$

$$\ln Q - \ln Q_0 = -kt$$

$$\ln\frac{Q}{Q_0} = -kt \Rightarrow \frac{Q}{Q_0} = e^{-kt}$$

Using rules of logs.

$Q = Q_0 e^{-kt}$ is the solution of the differential equation.

$Q_0 = 0.001$ and $Q = 0.0005$ when $t = 10$

So $0.0005 = 0.001\,e^{-10k} \Rightarrow e^{10k} = 2$

$$10k = \ln 2 \Rightarrow k = \tfrac{1}{10}\ln 2$$

Using rules of logs.

When $t = 20$, $Q = 0.001\,e^{-\frac{20}{10}\ln 2} = 0.001\,e^{-\ln 4}$

$$= 0.001\,e^{\ln\frac{1}{4}} = \frac{0.001}{4}$$

Using rules of logs.

$$= 0.000\,25$$

So $\qquad Q = 2.5 \times 10^{-4}$ when $t = 20$.

Revision exercise 4

In questions 1–10 integrate each expression.

1 $\cos \frac{2}{5}x$

2 $-\sin 4x$

3 e^{2-5x}

4 $\dfrac{x}{1+x^2}$

5 $\cos x \cos 2x$

6 $2x\,e^{3x}$

7 $\tan\left(x+\frac{\pi}{4}\right)$

8 $x\sec^2 x$

9 $\dfrac{6x^2}{x^3-1},\ x>1$

10 $x^2\ln x$

In questions 11–15 evaluate by integration:

11 $\displaystyle\int_1^3 \frac{1}{(x+2)(x+6)}\,\mathrm{d}x$

12 $\displaystyle\int_{\frac{1}{2}}^{\frac{3}{2}} x(4x^2+1)^2\,\mathrm{d}x$

13 $\displaystyle\int_0^{0.1} x^2\,e^{x^3}\,\mathrm{d}x$

14 $\displaystyle\int_0^{\frac{\pi}{4}} \cos^3 x\,\mathrm{d}x$

15 $\displaystyle\int_{\frac{\pi}{6}}^{\frac{\pi}{4}} \sin 2x \sin 3x\,\mathrm{d}x$

16 Find, in terms of e and π, the volume generated when the region bounded by the lines $x=1$, $x=3$ and $y=0$ and the curve with equation $y = e^{\frac{3x}{2}}$ is rotated completely about the x-axis.

17 The region bounded by the positive x- and y-axes, the line $y=2$ and the curve with equation $y=\ln x$ is rotated completely about the y-axis. Find the volume generated.

18 The region R is bounded by the x-axis, the line $x=\frac{2}{3}\pi$ and the curve with equation $y=\sin 2x$, as shown in the diagram.

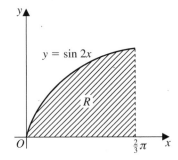

(a) Find the area of R.

(b) The region R is rotated through 2π radians about the x-axis to form a solid S. Find the volume of S.

19 The region R is bounded by the curve with equation $y = x^2 e^x$, the x-axis and the line $x = 1$.
Find, in terms of e, the area of R.

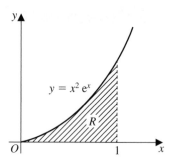

20 Prove that $\displaystyle\int_3^4 \frac{x^2 + 4}{x^2 - 4}\, dx = 1 + 2\ln\frac{5}{3}$.

21 Prove that $\displaystyle\int_0^{\frac{\pi}{2}} \frac{\cos x}{5 - 3\sin x}\, dx = \frac{1}{3}\ln\frac{5}{2}$.

22 Show that $\displaystyle\int \sec^4 x\, dx = \tan x + \frac{1}{3}\tan^3 x + C$, and hence

evaluate $\displaystyle\int_0^{\frac{\pi}{3}} \sec^4 x\, dx$ to 2 decimal places.

23 Find $\displaystyle\int \sin^2 x \cos^3 x\, dx$ in terms of $\sin x$.

24 Evaluate $\displaystyle\int_1^2 x\ln x\, dx$.

25 Find y in terms of x given that
$$\frac{dy}{dx} = (1 + 2\cos x)\cos 2x$$
and that $y = 0$ at $x = 0$.

26 Prove that $\displaystyle\int \frac{1}{\sin 2x}\, dx = \frac{1}{2}\ln|\tan x| + C$.

Hence solve the differential equation
$$\sin 2x \frac{dy}{dx} - y = 1$$
given that $y = 1$ at $x = \frac{\pi}{4}$.

27 Solve the differential equation $\dfrac{dy}{dx} = (3x^2 + 1)e^{-y}$, $x \geqslant 0$, given
that $y = 0$ at $x = 0$.
Hence find the value of y at $x = 2$.

28 Solve the differential equation $\dfrac{dy}{dx} = 2xy$.
Sketch the particular solution curve which passes through $(1, 3)$.

29 The following assertions are made about $f(x) \equiv x^2 \sin x$.

$$\mathbf{A} \int_0^\pi f(x)\,dx = 0 \qquad \mathbf{B} \int_0^{-t} f(x)\,dx = \int_0^t f(x)\,dx \qquad \mathbf{C} \int_{-t}^t f(x)\,dx = 0$$

If a statement is true prove it, if untrue give a counter-example.

30 For the differential equation $\dfrac{dy}{dx} = \dfrac{y+2}{x}$, $x \neq 0$ prove that

(a) $y = x - 2$ is the solution for $x = 2$ at $y = 0$,

(b) $y = 2x - 2$ is the solution line through $(4, 6)$.

Test yourself	**What to review**

If your answer is incorrect:

1 Find

(a) $\displaystyle\int (4 - 3x)^{-2}\,dx,$

(b) $\displaystyle\int e^{4-3x}\,dx.$

Review Heinemann Book P3 pages 105–107

---

2 Find $\displaystyle\int \cot^2 2x\,dx.$

Review Heinemann Book P3 pages 108–109

---

3 Find $\displaystyle\int x\,e^{-3x}\,dx.$

Review Heinemann Book P3 page 115–117

---

4 Use the substitution $v = x^2 + 4$ to evaluate $\displaystyle\int_1^2 \frac{x}{(x^2 + 4)^2}\,dx.$

Review Heinemann Book P3 pages 110–113

---

5 The finite region R is bounded by the x-axis, the lines $x = \dfrac{2\pi}{3}$ and $x = \pi$ and part of the curve with equation $y = \sin\frac{1}{2}x$.

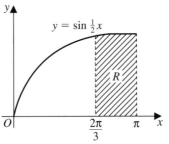

Review Heinemann Book P3 pages 121–126

about the x-axis.

(a) Find the area of R.

(b) Find the volume generated when R is rotated completely

6 Find the value of $\displaystyle\int_1^5 \frac{x}{(2x-1)^{\frac{3}{2}}}\,dx$.

Review Heinemann Book P3 pages 110–113

---

7 For a certain curve $\dfrac{dy}{dx} = \dfrac{4y}{3x-1}$ and the point $P(2, 1)$ lies on the curve.

Review Heinemann Book P3 pages 51–52 and 129–136

(a) Find the value of $\dfrac{d^2y}{dx^2}$ at P.

(b) Express y in terms of x.

---

8 An ellipse has the parametric equations

$$x = 5\cos t, \quad y = 2\sin t, \quad 0 \leqslant t < 2\pi$$

Review Heinemann Book P3 pages 125–126

(a) Sketch this ellipse.
(b) Use integration to find the area enclosed by the ellipse.

---

9 Prove that $\displaystyle\int \ln x\,dx = x(\ln x - 1) + C$, where C is a constant.

Hence find $\displaystyle\int (\ln x)^2\,dx$.

Review Heinemann Book P3 pages 115–117

---

10 (a) Find $\displaystyle\int \frac{\cos x}{4 - \sin^2 x}\,dx$.

Review Heinemann Book P3 pages 118–119

(b) Evaluate $\displaystyle\int_0^{\frac{\pi}{3}} \cos^3 x\,dx$.

Test yourself answers

1 (a) $\frac{1}{3}(4-3x)^{-1} + C$ (b) $-\frac{1}{3}e^{4-3x} + C$ **2** $-x - \frac{1}{2}\cot 2x + C$ **3** $-\frac{1}{9}e^{-3x}(3x+1) + C$ **4** $\frac{3}{80}$ **5** (a) 1 (b) $\frac{\pi}{12}(2\pi + 3\sqrt{3})$

6 $\frac{4}{3}$ **7** (a) $\frac{4}{25}$ (b) $y = \left(\dfrac{3x-1}{5}\right)^{\frac{4}{3}}$ **8** (b) 10π **9** $x(\ln x)^2 - 2x\ln x + 2x + C_0$ **10** (a) $\frac{1}{4}\ln\left(\dfrac{2+\sin x}{2-\sin x}\right) + C$ (b) $\dfrac{3\sqrt{3}}{8}$

Vectors

5

Key points to remember

1 A vector is a quantity which has both magnitude and direction in space.

2 A unit vector is a vector whose magnitude (modulus) is 1.

3 If **a** and **b** are parallel then $\mathbf{a} = \lambda\mathbf{b}$ for some scalar λ.

4 Vectors are added by the triangle law:

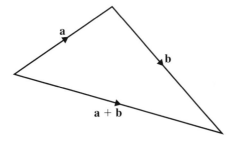

They can also be added or subtracted by the parallelogram law:

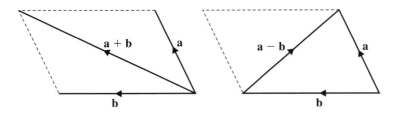

5 The vector $-\mathbf{a}$ has the same magnitude but opposite direction to **a**.

6 If **a** and **b** are non-parallel vectors and

$$\lambda\mathbf{a} + \mu\mathbf{b} = \alpha\mathbf{a} + \beta\mathbf{b}$$

where λ, μ, α, β are scalars, then

$$\lambda = \alpha \text{ and } \mu = \beta$$

7 If the points A and B have position vectors $\mathbf{a}$ and $\mathbf{b}$ with respect to an origin O,

then $\overrightarrow{AB} = \overrightarrow{OB} - \overrightarrow{OA} = \mathbf{b} - \mathbf{a}$,

and $\overrightarrow{OM} = \frac{1}{2}(\mathbf{a} + \mathbf{b})$,

where M is the mid-point of AB.

8 The distance between the points with coordinates (x_1, y_1, z_1) and (x_2, y_2, z_2) is
$$\sqrt{[(x_1 - x_2)^2 + (y_1 - y_2)^2 + (z_1 - z_2)^2]}.$$

9 The modulus of $\mathbf{a} = x\mathbf{i} + y\mathbf{j} + z\mathbf{k}$ is
$$|\mathbf{a}| = \sqrt{(x^2 + y^2 + z^2)}$$

10 If P has coordinates (x, y, z) then a unit vector in the direction $\overrightarrow{OP}$, where O is the origin, is
$$\frac{x\mathbf{i} + y\mathbf{j} + z\mathbf{k}}{\sqrt{(x^2 + y^2 + z^2)}}$$

11 The scalar (or dot) product of $\mathbf{a} = x_1\mathbf{i} + y_1\mathbf{j} + z_1\mathbf{k}$ and $\mathbf{b} = x_2\mathbf{i} + y_2\mathbf{j} + z_2\mathbf{k}$ is
$$\mathbf{a} \cdot \mathbf{b} = x_1 x_2 + y_1 y_2 + z_1 z_2 = |\mathbf{a}||\mathbf{b}| \cos \theta$$

12 The non-zero vectors $\mathbf{a}$ and $\mathbf{b}$ are perpendicular *if and only if* $\mathbf{a} \cdot \mathbf{b} = 0$.

13 The length of the projection of $\mathbf{a}$ on $\mathbf{b}$ is $\dfrac{\mathbf{a} \cdot \mathbf{b}}{|\mathbf{b}|}$

14 The vector equation of the straight line passing through the point with position vector $\mathbf{a}$ and parallel to the vector $\mathbf{b}$ is
$$\mathbf{r} = \mathbf{a} + \lambda\mathbf{b}$$

where λ is a scalar.

15 The straight line passing through the points A and B with position vectors $\mathbf{a}$ and $\mathbf{b}$ respectively has vector equation
$$\mathbf{r} = \mathbf{a} + \lambda(\mathbf{b} - \mathbf{a})$$

16 Cartesian equations of the straight line passing through the point (a_1, a_2, a_3) and parallel to the vector $b_1\mathbf{i} + b_2\mathbf{j} + b_3\mathbf{k}$ are
$$\frac{x - a_1}{b_1} = \frac{y - a_2}{b_2} = \frac{z - a_3}{b_3}$$

17 The straight line passing through (x_1, y_1, z_1) and (x_2, y_2, z_2) has cartesian equations
$$\frac{x - x_1}{x_2 - x_1} = \frac{y - y_1}{y_2 - y_1} = \frac{z - z_1}{z_2 - z_1}$$

Example 1

In the diagram

$\overrightarrow{AB} = \mathbf{b}$, $\overrightarrow{BC} = 2\mathbf{a}$ and

$\overrightarrow{AD} = \mathbf{b} + \frac{8}{3}\mathbf{a}$.

(a) Express $\overrightarrow{AC}$ in terms of $\mathbf{a}$ and $\mathbf{b}$.
(b) Prove that B, C and D are collinear.

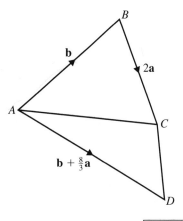

Answer

(a) $\overrightarrow{AC} = \overrightarrow{AB} + \overrightarrow{BC}$ Using 4

      $= \mathbf{b} + 2\mathbf{a}$

(b) $\overrightarrow{AC} + \overrightarrow{CD} = \overrightarrow{AD}$ Using 4

    $\mathbf{b} + 2\mathbf{a} + \overrightarrow{CD} = \mathbf{b} + \frac{8}{3}\mathbf{a}$ Using result from (a).

          $\overrightarrow{CD} = \frac{2}{3}\mathbf{a}$

Since $\overrightarrow{BC} = 2\mathbf{a}$ and $\overrightarrow{CD} = \frac{2}{3}\mathbf{a}$ are both parallel to vector $\mathbf{a}$ and the point C is common, the points B, C and D are collinear.   Using 3

Example 2

Given that $\mathbf{p} = 3\mathbf{i} - 2\mathbf{j} + 4\mathbf{k}$ and $\mathbf{q} = 5\mathbf{i} - \mathbf{j} + 2\mathbf{k}$, find
(a) $|\mathbf{q} - \mathbf{p}|$,
(b) a unit vector in the direction $\mathbf{q} - \mathbf{p}$.

Answer

(a) $\mathbf{q} - \mathbf{p} = (5\mathbf{i} - \mathbf{j} + 2\mathbf{k}) - (3\mathbf{i} - 2\mathbf{j} + 4\mathbf{k})$ Use brackets.

         $= 5\mathbf{i} - \mathbf{j} + 2\mathbf{k} - 3\mathbf{i} + 2\mathbf{j} - 4\mathbf{k}$ Simplify

         $= 2\mathbf{i} + \mathbf{j} - 2\mathbf{k}$ Collecting like terms.

$|\mathbf{q} - \mathbf{p}| = |2\mathbf{i} + \mathbf{j} - 2\mathbf{k}|$

         $= \sqrt{[2^2 + 1^2 + (-2)^2]}$ Using 9

         $= \sqrt{(4 + 1 + 4)}$

         $= \sqrt{9} = 3$

(b) A unit vector in the direction of $\mathbf{q} - \mathbf{p}$ is

    $\dfrac{\mathbf{q} - \mathbf{p}}{|\mathbf{q} - \mathbf{p}|} = \dfrac{2\mathbf{i} + \mathbf{j} - 2\mathbf{k}}{3}$ Using 10

         $= \frac{2}{3}\mathbf{i} + \frac{1}{3}\mathbf{j} - \frac{2}{3}\mathbf{k}$

Example 3
The vectors $\lambda\mathbf{i} + 2\mathbf{j} + 5\lambda\mathbf{k}$ and $3\mathbf{i} - 4\mathbf{j} + \lambda\mathbf{k}$ are perpendicular.
Find the possible values of λ.

Answer

If the vectors are perpendicular, their scalar product is zero.

Using **12**

That is $(\lambda\mathbf{i} + 2\mathbf{j} + 5\lambda\mathbf{k}).(3\mathbf{i} - 4\mathbf{j} + \lambda\mathbf{k}) = 0$

Using **11**

$$\Rightarrow 3\lambda - 8 + 5\lambda^2 = 0$$

But $5\lambda^2 + 3\lambda - 8 \equiv (5\lambda + 8)(\lambda - 1)$

Factorising.

So $\lambda = -\frac{8}{5}$ or $\lambda = 1$.

Example 4
The lines l_1 and l_2 have equations
l_1: $\mathbf{r} = (-2\mathbf{i} + 3\mathbf{j} + \mathbf{k}) + s(8\mathbf{i} + \mathbf{j} - 4\mathbf{k})$,
l_2: $\mathbf{r} = (\mathbf{i} + 2\mathbf{j} + 4\mathbf{k}) + t(6\mathbf{i} + 2\mathbf{j} + 3\mathbf{k})$

where s and t are scalar parameters.
At the point P on l_1, $s = 1$ and at the point Q on l_2, $t = -1$. The
acute angle between the directions of l_1 and l_2 is α. Find
(a) the distance between P and Q,
(b) the value of $\cos\alpha$.

Answer

(a) When $s = 1$, $\mathbf{r} = -2\mathbf{i} + 3\mathbf{j} + \mathbf{k} + 8\mathbf{i} + \mathbf{j} - 4\mathbf{k}$
$$= 6\mathbf{i} + 4\mathbf{j} - 3\mathbf{k}$$
So coordinates of P are $(6, 4, -3)$
When $t = -1$, $\mathbf{r} = \mathbf{i} + 2\mathbf{j} + 4\mathbf{k} - 6\mathbf{i} - 2\mathbf{j} - 3\mathbf{k}$
$$= -5\mathbf{i} + \mathbf{k}$$
So coordinates of Q are $(-5, 0, 1)$.
$PQ^2 = [6-(-5)]^2 + (4 - 0)^2 + (-3 - 1)^2$

Using **8**

$$= 11^2 + 4^2 + 4^2$$
$$= 153$$

The distance between P and Q is 12.4 units (3 s.f.).
(b) The direction of l_1 is parallel to vector $8\mathbf{i} + \mathbf{j} - 4\mathbf{k}$.

Using **14**

The direction of l_2 is parallel to vector $6\mathbf{i} + 2\mathbf{j} + 3\mathbf{k}$.

$$\cos\alpha = \frac{(8\mathbf{i} + \mathbf{j} - 4\mathbf{k}).(6\mathbf{i} + 2\mathbf{j} + 3\mathbf{k})}{|8\mathbf{i} + \mathbf{j} - 4\mathbf{k}||6\mathbf{i} + 2\mathbf{j} + 3\mathbf{k}|}$$

Using **11**

$$= \frac{48 + 2 - 12}{\sqrt{(8^2 + 1^2 + 4^2)}\sqrt{(6^2 + 2^2 + 3^2)}}$$

Using **11**

Using **9**

$$= \frac{38}{\sqrt{81}\sqrt{49}}$$

$$= \frac{38}{63}$$

Example 5

Referred to an origin O, the points A and B are given by $\overrightarrow{OA} = 5\mathbf{i} - \mathbf{j} + 7\mathbf{k}$ and $\overrightarrow{OB} = 6\mathbf{i} - 3\mathbf{j} - 5\mathbf{k}$.
Find

(a) a vector equation,
(b) parametric equations,
(c) cartesian equations,

of the line through A and B.

Answer

(a) $\overrightarrow{AB} = \overrightarrow{OB} - \overrightarrow{OA} = 6\mathbf{i} - 3\mathbf{j} - 5\mathbf{k} - (5\mathbf{i} - \mathbf{j} + 7\mathbf{k})$ Using **7**

$\qquad = \mathbf{i} - 2\mathbf{j} - 12\mathbf{k}$

That is, the line AB is parallel to $\mathbf{i} - 2\mathbf{j} - 12\mathbf{k}$ and, as the point A lies on AB, an equation of the line passing through A and B is

$$\mathbf{r} = 5\mathbf{i} - \mathbf{j} + 7\mathbf{k} + t(\mathbf{i} - 2\mathbf{j} - 12\mathbf{k})$$ Using **15**

where t is a scalar parameter.

(b) Take (x, y, z) as any point on the line $\mathbf{r} = x\mathbf{i} + y\mathbf{j} + z\mathbf{k}$ and you have

$$x\mathbf{i} + y\mathbf{j} + z\mathbf{k} = (5 + t)\mathbf{i} + (-1 - 2t)\mathbf{j} + (7 - 12t)\mathbf{k}.$$ See Book P3 page 161.

Equating coefficients of $\mathbf{i}$, $\mathbf{j}$ and $\mathbf{k}$

$$x = 5 + t, \quad y = -1 - 2t, \quad z = 7 - 12t$$ Using **6**

and these are parametric equations of the line.

(c) Notice that $t = x - 5$, $t = -\dfrac{y + 1}{2}$, $t = -\dfrac{z - 7}{12}$.

So $5 - x = \dfrac{y + 1}{2} = \dfrac{z - 7}{12}$ and these are cartesian equations of the   Using **16**
line.

Revision exercise 5

The vectors $\mathbf{a}$ and $\mathbf{b}$ are the position vectors of points A and B respectively, relative to the origin O.

$$\mathbf{a} = \mathbf{i} + \mathbf{j} + \mathbf{k}, \quad \mathbf{b} = 4\mathbf{i} + 5\mathbf{j} + \mathbf{k}$$

1 Find $\overrightarrow{AB}$.

2 Find a unit vector in the direction of $\overrightarrow{AB}$.

3 Find $\mathbf{a} \cdot \mathbf{b}$

4 Calculate $\angle AOB$ to the nearest degree.

5 Find a vector equation of the line passing through A and B.

For questions 6–10 repeat questions 1–5 for

$$\mathbf{a} = -4\mathbf{i} + 3\mathbf{j} - 5\mathbf{k}, \quad \mathbf{b} = -\mathbf{i} - \mathbf{j} + 7\mathbf{k}$$

For questions 11–15 repeat questions 1–5 for

$$\mathbf{a} = -2\mathbf{j} + 2\mathbf{k}, \quad \mathbf{b} = 4\mathbf{i} + 2\mathbf{j}$$

For questions 16–20 repeat questions 1–5 for

$$\mathbf{a} = 8\mathbf{i} - 5\mathbf{j} + 6\mathbf{k}, \quad \mathbf{b} = 2\mathbf{i} + 3\mathbf{j} - 18\mathbf{k}$$

21 Given that the vectors $\mathbf{p}$ and $\mathbf{q}$ are perpendicular, find the value of λ when

   (a) $\mathbf{p} = 2\mathbf{i} - 4\mathbf{j} + 3\mathbf{k}, \quad \mathbf{q} = 5\mathbf{i} + \mathbf{j} + \lambda\mathbf{k}$

   (b) $\mathbf{p} = 3\mathbf{i} + 7\mathbf{j} + \lambda\mathbf{k}, \quad \mathbf{q} = 8\mathbf{i} - 7\mathbf{j} + \lambda\mathbf{k}$

   (c) $\mathbf{p} = -3\lambda\mathbf{i} + 2\lambda\mathbf{j} + 2\mathbf{k}, \quad \mathbf{q} = \lambda\mathbf{i} + 4\mathbf{j} - 2\mathbf{k}$

22 In $\triangle ABC$, L and M are the mid-points of BC and AC. Prove that

   (a) $2\overrightarrow{LM} = \overrightarrow{BA}$

   (b) area of $\triangle ABC = 4 \times$ area of $\triangle CLM$.

23 Prove, using vectors, that the diagonals of a parallelogram bisect each other.

24 The lines l_1 and l_2 have equations

   l_1: $\mathbf{r} = -2\mathbf{i} + 3\mathbf{j} + 4\mathbf{k} + s(8\mathbf{i} + 4\mathbf{j} - 4\mathbf{k})$,

   l_2: $\mathbf{r} = -3\mathbf{i} + 10\mathbf{j} + 6\mathbf{k} + t(3\mathbf{i} - \mathbf{j} - 2\mathbf{k})$,

   where s and t are scalar parameters.

   (a) Prove that the lines l_1 and l_2 intersect.

   (b) Find the coordinates of their point of intersection.

25 Find the acute angle between the lines l_1 and l_2 in question 24.

26

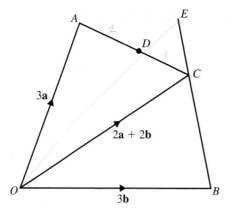

In the diagram, $\overrightarrow{OA} = 3\mathbf{a}$, $\overrightarrow{OB} = 3\mathbf{b}$, and $\overrightarrow{OC} = 2\mathbf{a} + 2\mathbf{b}$. The point D is on AC such that $AD:DC = 2:1$ and E is the point on BC produced such that $\overrightarrow{BE} = k\overrightarrow{BC}$, where k is constant.

(a) Express, in terms of **a** and **b**,

(i) $\overrightarrow{BC}$,   (ii) $\overrightarrow{AC}$,   (iii) $\overrightarrow{OD}$.

(b) Prove that $\overrightarrow{OE} = 2k\mathbf{a} + (3 - k)\mathbf{b}$.

Given also that ODE is a straight line, find

(c) the value of k,

(d) $\overrightarrow{OE}$, in terms of **a** and **b** only,

(e) the ratio $OD : DE$.   [E]

27   Relative to an origin O, $\overrightarrow{OA} = \mathbf{i} + 3\mathbf{j} + 4\mathbf{k}$ and
$\overrightarrow{OB} = 2\mathbf{i} + \mathbf{j} + 6\mathbf{k}$. The points C and D have coordinates
$(2, 2, 2)$ and $(4, 6, 6)$ respectively.

(a) Find, to the nearest degree, the acute angle between the directions of the lines AB and CD.

(b) Prove that, in fact, the lines AB and CD do not intersect.

28   Referred to an origin O, the points A, B, C, D have coordinates

$$(-1, 12, 12), \quad (1, 8, 8), \quad (2, 9, 12), \quad (-2, 11, 8)$$

respectively. Prove that A, B, C, D are the vertices of a square.

29   Find the shortest distance of the point $(6, 6, -1)$ from the line with equation

$$\mathbf{r} = 2\mathbf{i} + \mathbf{j} - 3\mathbf{k} + t(\mathbf{i} + 2\mathbf{j} - \mathbf{k}).$$

30   Referred to the origin O, the points A, B, C, D are such that
$\overrightarrow{OA} = \mathbf{i} + 3\mathbf{j} + 5\mathbf{k}$, $\overrightarrow{OB} = 2\mathbf{i} + 5\mathbf{j} + 3\mathbf{k}$,
$\overrightarrow{OC} = -\mathbf{i} + \mathbf{j} - 9\mathbf{k}$, $\overrightarrow{OD} = \alpha\mathbf{i} + \beta\mathbf{j} + \gamma\mathbf{k}$.
Find

(a) $\overrightarrow{BA}$ and $\overrightarrow{BC}$,

(b) $\angle ABC$ to the nearest $0.1°$

(c) the values of α, β and γ if $ABCD$ is a parallelogram,

(d) the position vector of the point E on the line AC which is such that $\overrightarrow{AE} = 3\overrightarrow{EC}$.

Test yourself	What to review

1 In the diagram,

$\overrightarrow{AB} = \mathbf{p}$, $\overrightarrow{BD} = \mathbf{q}$ and $\overrightarrow{BC} = \mathbf{r}$

Find $\overrightarrow{AD}$ and $\overrightarrow{CD}$.

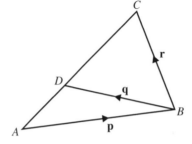

Review Heinemann Book P3 pages 141–144

2 The vectors **a** and **b** are not parallel.
Given that

$$(\lambda - 2\mu)\mathbf{a} + (\lambda - 1)\mathbf{b} = 8\mathbf{a} + (5 + \mu)\mathbf{b},$$

find the values of the scalars λ and μ.

Review Heinemann Book P3 pages 147–148

3 Find a unit vector in the direction of $24\mathbf{i} - 7\mathbf{j}$.

Review Heinemann Book P3 pages 154–155

4 Given that the unit of distance is cm, find the distance between the points $A(13, -5, 7)$ and $B(7, -7, 10)$.

Review Heinemann Book P3 pages 158–160

5 The vectors **u** and **v** are given by

$$\mathbf{u} = 4\mathbf{i} + 5\mathbf{j} - 2\mathbf{k} \text{ and } \mathbf{v} = 3\mathbf{i} - 2\mathbf{j} + \mathbf{k}.$$

(a) Find $\mathbf{u} . \mathbf{v}$
(b) What deduction can you make from your answer to part (a)?

Review Heinemann Book P3 pages 164–165

6 Find in vector form an equation of the line passing through $A(-3, -2, 4)$ and $B(4, -6, -5)$.

Review Heinemann Book P3 pages 168–172

7 The lines l_1 and l_2 are given by

$$l_1: \mathbf{r} = \begin{pmatrix} -2 \\ 3 \\ 4 \end{pmatrix} + s \begin{pmatrix} 2 \\ 1 \\ -1 \end{pmatrix}$$

$$l_2: \mathbf{r} = \begin{pmatrix} -3 \\ 10 \\ 5 \end{pmatrix} + t \begin{pmatrix} 3 \\ -1 \\ -2 \end{pmatrix}$$

where s and t are scalar parameters.
Prove that these lines do **not** meet.

Review Heinemann Book P3 pages 168–172

---

8 Find the length of the projection of the vector $3\mathbf{i} - 2\mathbf{j} + 5\mathbf{k}$ in the direction of the vector $\mathbf{i} - 2\mathbf{j} + 2\mathbf{k}$.

Review Heinemann Book P3 pages 165–166

---

9 Referred to an origin O, the points A and B are given by

$$\overrightarrow{OA} = 4\mathbf{i} - 5\mathbf{j} + 2\mathbf{k} \text{ and } \overrightarrow{OB} = -\mathbf{i} + 2\mathbf{j} - 3\mathbf{k}$$

Find $\angle AOB$ to the nearest degree.

Review Heinemann Book P3 pages 164–165

---

10 In $\triangle OAB$, $\overrightarrow{OA} = 2\mathbf{a}$ and $\overrightarrow{OB} = 2\mathbf{b}$.
The mid-points of OA and OB are L and M respectively.
Prove that $2\overrightarrow{LM} = \overrightarrow{AB}$ and state, in words, what this result implies.

Review Heinemann Book P3 pages 141–148

Test yourself answers

1 $\overrightarrow{AD} = \mathbf{p} + \mathbf{q}$, $\overrightarrow{CD} = \mathbf{q} - \mathbf{r}$ **2** $\lambda = 4$, $\mu = -2$ **3** $\frac{24}{25}\mathbf{i} - \frac{7}{25}\mathbf{j}$ **4** 7 cm   **5 (a)** 0 **(b)** The vectors are perpendicular

6 $\mathbf{r} = 4\mathbf{i} - 6\mathbf{j} - 5\mathbf{k} + t(7\mathbf{i} - 4\mathbf{j} - 9\mathbf{k})$ **8** $\frac{17}{3}$ **9** 143°   **10** LM is parallel to AB and $LM = \frac{1}{2}AB$.

Examination style paper

Answer all questions   **Time 90 minutes**

Use only a basic scientific calculator when answering this paper.

1. Find the remainder when $(x + 1)^3 + x^4$ is divided by $x + 3$.
   (3 marks)

2. (a) Show in a sketch the finite region R which is bounded by the positive x-axis, the y-axis and part of the curve with equation $y = 1 + \sin 2x$. **(2 marks)**

   (b) Find the area of R, giving your answer in terms of π.
   (3 marks)

3. A circle has equation $x^2 + y^2 - 8x + 2y + 8 = 0$. The point C on the circle is the nearest point to the origin O.
   Find the distance between O and C. **(5 marks)**

4. Solve the differential equation

   $$2y\cos^2 x \frac{dy}{dx} = 2\tan x + 1$$

   where $y = 2$ at $x = \frac{\pi}{4}$, giving your final answer in the form $y = f(x)$. **(7 marks)**

5. (a) Express $f(x) \equiv \dfrac{1}{(2x + 1)^2(x + 2)}$ in partial fractions.
   (4 marks)
   (b) Hence find the value of $\dfrac{dy}{dx}$ at $x = 0$. **(4 marks)**

6. $$f(x) \equiv (1 + 3x)^{-\frac{2}{3}} - (1 - 4x)^{\frac{1}{2}}$$

   (a) Expand $f(x)$ in ascending powers of x as far as the term in x^3.

   (7 marks)
   (b) State the set of values of x for which the expansion is valid. **(2 marks)**

7. At time t, the rate of increase in the concentration C of micro-organisms in a biological experiment is equal to λC, where λ is a positive constant. When $t = 0$, $C = C_0$.
    (a) By forming and solving a differential equation show that

    $$C = C_0\, e^{\lambda t}. \qquad \textbf{(7 marks)}$$

    (b) Find t in terms of λ when $C = \frac{3}{2}C_0$.   **(3 marks)**

8. *In each sketch below you should identify the coordinates of any points at which a curve meets the coordinate axes.*

    The curve C_1 has equation $y = (x - 1)(4 - x)$.
    (a) Sketch, in separate diagrams,
    (i) C_1,   **(3 marks)**
    (ii) the curve C_2 with equation $y = |(x - 1)(4 - x)|$,   **(3 marks)**

    (iii) the curve C_3 with equation $y = \dfrac{1}{(x - 1)(4 - x)}$.   **(3 marks)**

    (b) Find an equation of the normal at the point $P(t + 1,\ 3t - t^2)$ on the curve C_1.   **(4 marks)**

9. (a) Find, in vector form, an equation of the line passing through $A\,(-4,\ 13,\ 7)$ and $B\,(8,\ -11,\ -5)$.   **(3 marks)**

    Referred to an origin O, the point C has position vector $5\mathbf{i} - 4\mathbf{j} + 2\mathbf{k}$ and D is the point on the line AB such that CD is perpendicular to AB.
    (b) Find the coordinates of D and hence state the value of the ratio $AD : DB$.   **(8 marks)**

    (c) Calculate the size of $\angle OAB$, giving your answer to the nearest degree.   **(4 marks)**

Answers

Revision exercise 1

1 $\dfrac{2}{x-2} - \dfrac{4}{2x+1}$

2 $\dfrac{1}{2}\left[\dfrac{1}{4x-3} - \dfrac{1}{4x-1}\right]$

3 $\dfrac{-\frac{1}{4}}{x+1} + \dfrac{\frac{1}{5}}{x+2} + \dfrac{\frac{1}{20}}{x-3}$

4 $2 + \dfrac{12}{x-3}$

5 $\dfrac{3}{x-1} - \dfrac{3}{x+1} + \dfrac{2}{(x+1)^2}$

6 $\dfrac{6x+1}{2x^2+1} - \dfrac{3}{x+3}$

7 $1 + \dfrac{12}{x-3} + \dfrac{36}{(x-3)^2}$

8 $3x - 2 - \dfrac{3}{2x-1} + \dfrac{1}{x+4}$

9 6

10 123　　**11** -61　　**12** -5　　**13** 56

14 $\frac{27}{8}$　　**15** $1 + 6x + 24x^2 + 80x^3$; $|x| < \frac{1}{2}$

16 $1 + 3x - \frac{3}{2}x^2 + \frac{5}{2}x^3$; $|x| < \frac{1}{4}$

17 $\frac{1}{4}\left[1 - \frac{3}{4}x + \frac{9}{16}x^2 - \frac{27}{64}x^3\right]$; $|x| < \frac{4}{3}$

18 $1 - \frac{1}{2}x - \frac{1}{2}x^2 + \frac{1}{4}x^3$; $|x| < 1$

19 $1 - 3x + \frac{9}{2}x^2 - \frac{27}{2}x^3$; $|x| < \frac{1}{3}$

20 $1 + 18x - 39x^2 + 172x^3$; $|x| < \frac{1}{2}$

21 45　　**22** -3

23 $\dfrac{1}{(1+x)^2} - \dfrac{2}{1+x} + \dfrac{1}{1-2x}$; $2x + 5x^2$; $|x| < \frac{1}{2}$

25 $\dfrac{1}{\sqrt{5}}\left[-\frac{1}{10}x + \frac{3}{200}x^2\right]$; 5.916

26 $p = -13$, $q = -26$; -69

28 $\dfrac{1}{x^3} + \dfrac{5}{x^4} + \dfrac{14}{x^5}$

Revision exercise 2

1 $-\frac{5}{2}\sin\frac{5}{2}x$　　**2** $-4\tan 2x \sec^2 2x$

3 $(3x-1)^{-\frac{2}{3}}$　　**4** $\dfrac{4}{4x+3}$

5 $\dfrac{2x + 10x^2}{\sqrt{(4x+1)}}$　　**6** $\frac{1}{2}e^{\frac{1}{2}x}\sin 3x + 3e^{\frac{1}{2}x}\cos 3x$

7 $\ln(\sec x + \tan x) + x \sec x$

8 $\dfrac{4x + 9x^2 - x^4}{(x^3+2)^2}$

9 $\dfrac{2x\,e^{x^2}\tan 3x - 3e^{x^2}\sec^2 3x}{\tan^2 3x}$　　**10** $\dfrac{2\sin x}{\cos^3 x}$

11 $\operatorname{cosec} x$　　**12** $\dfrac{16x}{x^4 - 16}$

13 $-\frac{7}{4}\cot t$　　**14** $\dfrac{\sec t(1 + \tan t)}{\tan t + \sec^2 t}$　　**15** $\dfrac{4}{3t}$

16 $\dfrac{\cos t + \cos 2t}{1 + \sin 2t}$

17 $\dfrac{dy}{dx} = -\dfrac{x^2}{y^2}$; $y - 3 = -\frac{4}{9}(x-2)$

18 $\dfrac{dy}{dx} = \dfrac{3x+y}{y-x}$; $y - 3 = 3(x-1)$

19 $\dfrac{dy}{dx} = \dfrac{x-7}{1-y}$; $y - 2 = 3(x-4)$

20 $\dfrac{dy}{dx} = \dfrac{25 - 20x + 3x^2}{2y}$; $y + 2 = -2(x-1)$

21 $y - \frac{1}{2} = 2(x+1)$　　**22** $\left(\frac{121}{9}, -\frac{22}{3}\right)$

23 $y + 1 = -4\sqrt{3}\left(x - \dfrac{\sqrt{3}}{2}\right)$

24 $(0, 1)$ minimum; $(-4, \frac{1}{9})$ maximum

25 $2.2\ \text{cm s}^{-1}$

26 **(a)** 9.001　　**(b)** 0.008　　**(c)** 3.992

Revision exercise 3

1 **(a)** $x^2 + y^2 = 36$

　　(b) $x = 6\cos t$, $y = 6\sin t$, $0 \leqslant t < 2\pi$

2 **(a)** $(x-2)^2 + (y-1)^2 = 36$

　　(b) $x = 2 + 6\cos t$, $y = 1 + 6\sin t$, $0 \leqslant t < 2\pi$

3 **(a)** $(x+3)^2 + (y+5)^2 = 16$

　　(b) $x = 4\cos t - 3$, $y = 4\sin t - 5$, $0 \leqslant t < 2\pi$

4　(a)　$(x-5)^2+(y-4)^2=25$

　　(b)　$x=5+5\cos t,\ y=4+5\sin t,\ 0\leqslant t<2\pi$

5　$(x-5)^2+(y-6)^2=25$

6　$(x-2)^2+y^2=4$

7　$x^2+y^2-6x+8y-16=0$

8　$\left(\tfrac{7}{2},12\right);\ 12.5$　　9　$\left(\tfrac{5}{2},\tfrac{7}{2}\right);\ \tfrac{1}{2}$

10　$(3,5);\ 4$

11　$(4,1)$, radius $=3$; $(6.4,2.9)$ is outside the circle.

13　$xy=28$　　　　　14　$6y=x^2+54$

15　$25x=150y-y^2$　16　$x^2-y^2=1$

22　$(x+3)(x+6)+y(y-9)=0$

24　Curves intersect at $(8,2)$

25　(a)　$-\tfrac{5}{4}\leqslant y\leqslant 5$

Revision exercise 4

1　$\tfrac{5}{2}\sin\tfrac{2}{5}x+C$　　　2　$\tfrac{1}{4}\cos 4x+C$

3　$-\tfrac{1}{5}e^{2-5x}+C$　　　4　$\tfrac{1}{2}\ln(1+x^2)+C$

5　$\sin x-\tfrac{2}{3}\sin^3 x+C$　6　$\tfrac{2}{3}xe^{3x}-\tfrac{2}{9}e^{3x}+C$

7　$\ln\sec\left(x+\tfrac{\pi}{4}\right)+C$

8　$x\tan x+\ln\cos x+C$　　9　$2\ln(x^3-1)+C$

10　$\tfrac{1}{3}x^3\ln x-\tfrac{1}{9}x^3+C$　11　$\tfrac{1}{4}\ln\tfrac{35}{27}$

12　$41\tfrac{1}{3}$　　　　　　13　$\tfrac{1}{3}(e^{0.001}-1)$

14　$\tfrac{5}{12}\sqrt{2}$　　　　15　$\tfrac{3}{10}\sqrt{2}-\tfrac{1}{5}$

16　$\dfrac{\pi e^3}{3}(e^6-1)$　　　17　$\tfrac{\pi}{2}(e^4-1)$

18　(a)　$\tfrac{3}{4}$　(b)　$\tfrac{\pi}{48}(16\pi-3\sqrt{3})$

19　$e-2$　　　　　　22　3.46

23　$\tfrac{1}{3}\sin^3 x-\tfrac{1}{5}\sin^5 x+C$

24　$2\ln 2-\tfrac{3}{4}$

25　$y=\tfrac{1}{3}\sin 3x+\tfrac{1}{2}\sin 2x+\sin x$

26　$(y+1)^2=4\tan x$

27　$y=\ln(x^3+x+1);\ \ln 11$

28　$y=Ae^{x^2}$; the solution curve through $(1,3)$ is $y=3e^{x^2-1}$

29　Only C is true (Note $x^2\sin x$ is an odd function.)

Revision exercise 5

1　$3i+4j$　　　　2　$\tfrac{1}{5}(3i+4j)$　3　10

4　$27°$　　　　　5　$r=(i+j+k)+t(3i+4j)$

6　$3i-4j+12k$　7　$\tfrac{1}{13}(3i-4j+12k)$

8　-34　　　　　9　$132°$

10　$r=(-i-j+7k)+t(3i-4j+12k)$

11　$4i+4j-2k$　12　$\tfrac{2}{3}i+\tfrac{2}{3}j-\tfrac{1}{3}k$

13　-4　　　　　14　$108°$

15　$r=-2j+2k+t(2i+2j-k)$

16　$-6i+8j-24k$

17　$\tfrac{1}{13}(-3i+4j-12k)$

18　-107　　　　19　$121°$

20　$r=2i+3j-18k+t(3i-4j+12k)$

21　(a)　-2　(b)　±5　(c)　2 or $\tfrac{2}{3}$

24　(b)　$(6,7,0)$　25　$40.2°$

26　(a)　(i) $2a-b$　(ii) $2b-a$　(iii) $\tfrac{7}{3}a+\tfrac{4}{3}b$

　　(c)　$\tfrac{7}{5}$　(d)　$\tfrac{14}{5}a+\tfrac{8}{5}b$　(e)　$5:1$

27　(a)　$84°$　　　29　$\sqrt{21}$ units of length

30　(a)　$\overrightarrow{BA}=-i-2j+2k,$
　　　　$\overrightarrow{BC}=-3i-4j-12k$

　　(b)　$109.5°$　(c)　$\alpha=-2,\ \beta=-1,\ \gamma=-7$

　　(d)　$-\tfrac{1}{2}i+1\tfrac{1}{2}j-5\tfrac{1}{2}k$

Examination style paper

1　73　　2　(b)　$\tfrac{1}{2}+\tfrac{3}{4}\pi$　　3　$\sqrt{17}-3$

4　$y=(\tan^2 x+\tan x+2)^{\frac{1}{2}}$

5　(a)　$\dfrac{1}{9}\left[\dfrac{6}{(1+2x)^2}-\dfrac{2}{1+2x}+\dfrac{1}{x+2}\right]$　(b)　$-\tfrac{9}{4}$

6　(a)　$7x^2-\tfrac{28}{3}x^3$　　(b)　$|x|<\dfrac{1}{4}$

7　(b)　$\lambda^{-1}\ln\tfrac{3}{2}$

8　(a)　(i)　$(0,-4),(1,0),(4,0)$

　　　　(ii)　$(0,4),(1,0),(4,0)$

　　　　(iii)　$\left(0,-\tfrac{1}{4}\right)$

　　(b)　$y-3t+t^2=\dfrac{-1}{(3-2t)}(x-t-1)$

9　(a)　$r=-4i+13j+7k+t(i-2j-k)$

　　(b)　$(4,-3,-1);\ 2:1$　(c)　$9°$